140字だから楽しく読める
英語 twitter 多読術

Enjoy English with Twitter!

足立恵子　ジョナサン・ナクト 著

> 多読したいけど、絵本じゃあ物足りない

> そうそう、でも洋書は難しいよね

> 会話調140字のTwitter形式なら、楽しく英文が読めるよ！

SANSHUSHA

はじめに

　ツイッターを使って英語を勉強したいと思っている人は多いと思います。ツイッターとは、インターネット上のコミュニケーション・サービスで、140字以内の短いつぶやき（tweet＝ツイート）を投稿、ほかの人を「フォロー」してその人のツイートを読んだり、人のツイートに返信したりすることができます。

　ツイッターを使えば、世界中の人のつぶやきを英語で読むことができます。でも、普段英語を読み慣れていない人は、最初からリアルな英語ツイッターの世界に入っていくことにちょっとためらいを感じるかもしれません。そこで本書では、立場や興味が異なる三人の日本人を登場させ、さまざまな国にいる外国人と日々ツイッターでやりとりする様子を、物語形式で紹介しています。旅行が好きな30代独身女性、会社の公用語が英語になってあせっているエンジニア、日韓のアイドル大好き！ のフリーターが登場。最初は自分に近い登場人物のツイートから読み始め、慣れてきたら、小説を読むようなつもりで、ほかの人のツイートも楽しんでみてください。「意外と簡単なことが書いてあるんだな」と思えるようになったら、ぜひ実際のツイッターの世界ものぞいてみて、フォローしたくなるような相手を見つけましょう。日々ツイッターを読むことで、英語に触れる時間が、自然に確保できるようになってくるはずです。

　このツイッターは、実は「多読」の素材として最適なのです。多読とは、辞書を引いたり文法的な意味合いを調べたりすることなく、大量の英語に目を通していくことで、意味を推測する力を養い、英語的な発想に親しんでいく方法です。

　ツイッターを多読学習にお勧めしたい理由としては、次のようなものがあります。

1. 短いので読みやすい

　140字以内という文字数制限があるので、あまり込み入った、難しい話は出てきません。ちょっとしたひとことを読み流していけばいいのです。

2. 英語がやさしい

　基本的には会話に近く、シンプルな分かりやすい文章で書かれています。

3. 興味・親近感を持ちやすい

　自分に近い興味を持つ人のツイートをフォローすれば、どんなことが書かれているのが気になり、読み続けるのが苦でなくなります。

　多読の素材としてはよく海外のベストセラー小説などが読めるペーパーバックが使われますが、初級者には少々難しいかもしれません。「やさしい絵本から始めれば」という人もいますが、「今さら絵本というのも……」と思われる方もいることでしょう。インターネット上で、いつでも誰でも無料で読むことができるこのツイッターを、ぜひ多読の素材として使ってみてください。そして、これまで読んだ表現を参考に、自分でも英語でツイートを書いてみたいと思うようになるかもしれません。

　私たち自身も、ツイッターを利用して世界のさまざまな人々とやりとりをすることで、同じ興味・関心を持つ人とふれあう機会がさらに増え、海外の事情をリアルタイムで知ることができるようになりました。本書の中で触れていますが、旅先で撮った写真をアップロードしたり、待ち合わせの際に利用したり、いろいろな場面でツイッターを活用し、英語力向上に役立てていただければと思います。

2011年3月
足立恵子
ジョナサン・ナクト

contents

p7
Story 1
A Trip to Canada
カナダ旅行つぶやき日記

p74
COLUMN 1　自分の意見や希望を伝える／依頼する表現

p75
Story 2
My Company's Official Language will be English!
おやじのグローバル化（T_T）泣き言つぶやき

p142
COLUMN 2　普段の会話の中で使える表現

p143
Story 3
We Love SuperBoys!
コンサートに行こう♪ワクワクつぶやき

Story 1
A Trip to Canada

カナダ旅行つぶやき日記

Marippe

Marippe
@mari_tabi Yokohama, Japan
女性。36歳独身。メーカー勤務

I love overseas trips! I've been to England, France, Germany, China, and Thailand... Want to meet new people!

海外旅行大好き！ イギリス、フランス、ドイツ、中国、タイに行ったことがあります……新しい人たちと会ってみたい！

Kathleen Taylor
@kathleen_tl Victoria, Canada
女性。ビクトリアに住むカナダ人

I love sunny days, pretty flowers, and my beautiful town. I'm always busy in my garden!

晴れた日ときれいな花と、私の美しい街が大好き。いつも庭の手入れで忙しくしています！

Patty Cooper
@patty_coop Victoria, Canada
女性。ビクトリアに住むカナダ人。
kathleenのガーデニング仲間

Sunday mornings, the smell of pie, hot coffee, and a great husband! These are my favorites.

日曜の朝、パイの香り、熱いコーヒー、そしてステキなだんなさま！ これが私の好きな物です。

Liu Fei
@sweet_pie01 Shanghai
女性。上海に住む中国人。mariとは境遇が似ていて互いに共感し合える

A Chinese business woman working hard everyday, always looking for a better life. Please let me know about things that are fun or interesting!

毎日一生懸命働いていて、いつもよりよい生活を求めている中国人女性です。楽しいこと、面白いことを教えてください！

英語 Twitter 多読術　**7**

Story 1   A Trip to Canada

DAY 1

mari_tabi

I'm wondering where to go next ... maybe Canada would be a good idea. I haven't been there yet.

> I'm wondering 〜：〜だろうかと思う

mari_tabi

I stopped by a travel agency to look at some brochures of Canadian tours. Vancouver seems quite interesting to me.

> stop by 〜：〜に寄る　　travel agency：旅行会社
> brochure：パンフレット　　quite：とても

kathleen_tl

@mari_tabi Hi from Canada! Visit Victoria too! It's my hometown and quite a beautiful city.

> hometown：住んでいる街、故郷

mari_tabi

@kathleen_tl Thanks for the info. I'd like to know more about your town. Any more info will be helpful.

> info：情報（informationの略）　　helpful：役に立つ、助かる

kathleen_tl

@mari_tabi There's a big garden called "Butchart Gardens" in Victoria. You should go there. It's especially nice in the spring or autumn.

> Butchart Gardens：ブッチャート・ガーデン（ビクトリアにある巨大な庭園。観光地として有名）　　especially：特に

A Trip to Canada **Story 1**

1日目

mari_tabi

次はどこへ行こうかな……カナダとかいいかも。まだ行ったことないし。

mari_tabi

旅行会社に寄って、カナダ・ツアーのパンフレットを見てみた。バンクーバーってとっても面白そう。

kathleen_tl

@mari_tabi カナダからこんにちは！ ビクトリアにも来て！ 私の住んでいる街で、とってもきれいな所なの。

mari_tabi

@kathleen_tl 情報をありがとう。あなたの街のこと、もっと知りたいな。何かもっと教えてくれたら助かります。

kathleen_tl

@mari_tabi ビクトリアには、「ブッチャート・ガーデン」という大きな庭園があるの。ぜひ行ってみて。春や秋には特にステキなの。

 **mari_tabi** Marippe
 **kathleen_tl** Kathleen Taylor
 patty_coop Patty Cooper
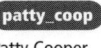  **sweet_pie01** Liu Fei

Story 1　A Trip to Canada

kathleen_tl

Today I planted rosemary in my garden. It needs warm dry weather. I hope the weather will be nice for a while.

plant：植える（名詞で「植物」）　　for a while：しばらくの間

mari_tabi

@kathleen_tl I like gardening too. I have a garden on my balcony and I'm taking care of 6 different plants. I'll see what grows the best.

take care：育てる、面倒を見る

sweet_pie01

@mari_tabi Hello from Shanghai! I had a tough day! I work too much and really need to take a vacation. Have you decided to go to Canada?

tough：大変な、キツい　　take a vacation：休暇を取る　　decide：決心する

DAY 2

mari_tabi

Just looked at some photos in a guidebook! Canada really seems like a nice place to visit. I just hope it's not too cold yet.

guidebook：ガイドブック　　seem like ～：～のように見える

kathleen_tl

@mari_tabi A little chilly at night, but don't worry about it. When it's sunny, it's rather warm during the daytime in Victoria.

chilly：冷える、寒い　　worry：心配する　　sunny：天気がいい、日が照っている
daytime：昼間、日中

kathleen_tl

今日はお庭にローズマリーを植えてみた。暖かくて乾燥した気候が必要なの。しばらく天気がいいといいけれど。

mari_tabi

@kathleen_tl 私もガーデニングが好き。バルコニーに庭があって、6種類の植物を育ててるの。どれが一番よく育つか見てるんだ。

sweet_pie01

@mari_tabi 上海からこんにちは！ 今日は大変だった！ 私ってば働きすぎ、本気で休暇を取る必要があるわね。カナダに行くの決めた？

2日目

---

mari_tabi

ガイドブックの写真見て！ カナダって本当に旅行するのによさそうな所。まだそんなに寒くないといいけれど。

kathleen_tl

@mari_tabi 夜はちょっと冷えるけど、心配しないで。天気がいいと、ビクトリアでは昼間はわりと暖かいの。

 mari_tabi Marippe **kathleen_tl** Kathleen Taylor **patty_coop** Patty Cooper  **sweet_pie01** Liu Fei

Story 1 A Trip to Canada

mari_tabi

@kathleen_tl Tell me more about Victoria! I heard there are a lot of historical buildings in that area.

> historical：歴史のある

kathleen_tl

@mari_tabi The Parliament Building is worth visiting. And just walking around town is a great experience.

> The Parliament Building：議事堂（ビクトリアの中心にあるブリティッシュ・コロンビア州議事堂のこと。parliamentは「議会」）

mari_tabi

@kathleen_tl I'd love to try afternoon tea. I didn't have time to do it when I went to London. Are there any good places to go?

kathleen_tl

Everyone goes to the Fairmont Empress Hotel. It was built in 19th century style. RT **@mari_tabi** I'd love to try afternoon tea

> Fairmont Empress Hotel：フェアモント・エンプレス・ホテル（ビクトリアの内湾に面した所にある高級ホテル）
> 19th century style：19世紀の様式（centuryは「世紀」）

kathleen_tl

@mari_tabi My boyfriend lives in Toronto, it's also a very interesting city to visit. Actually, he's been there for a month for work.

> actually：実は、実際は　　for work：仕事で

mari_tabi

Oh no! I'm quite busy at work now. Can I really take a vacation next month? I hope T-san can cover for me.

> cover：代わりをする

A Trip to Canada **Story 1**

mari_tabi
@kathleen_tl ビクトリアのこと、もっと教えて！ その地域には、歴史ある建物がたくさんあるって聞いたけれど。

kathleen_tl
@mari_tabi 議事堂は行く価値があるわね。それに街を歩いているだけでも、いい経験になるわよ。

mari_tabi
@kathleen_tl ぜひアフタヌーン・ティーを試してみたいな。ロンドンに行ったとき、やる暇がなかったの。どこかいい所ある？

kathleen_tl
みんなフェアモント・エンプレス・ホテルに行くわね。19世紀の様式で建てられたの。RT @mari_tabi ぜひアフタヌーン・ティーを試してみたいな。

kathleen_tl
@mari_tabi 私のボーイフレンドはトロントに住んでいて、そこも行ってみるととても面白い街なのよ。実は、彼、仕事でもうそこに1カ月いるの。

mari_tabi
ああ、どうしよう！ 今仕事ですごく忙しいの。本当に来月休みが取れるのかな？ Tさんが私の代わりをやってくれるといいんだけれど。

mari_tabi
Marippe

kathleen_tl
Kathleen Taylor

patty_coop
Patty Cooper

sweet_pie01
Liu Fei

Story 1 A Trip to Canada

DAY 3

mari_tabi

There was a problem with one of our company products and I spent the whole day explaining and apologizing to customers. I'm exhausted!

whole day：丸1日　　explain：説明する　　apologize：謝る
customer：お客様、顧客　　exhausted：疲れ切って

mari_tabi

People say, "you are the manager." Yes, I am but what's good about that? Nobody listens to me!

manager：マネジャー、責任者

kathleen_tl

@mari_tabi Don't get upset. When something bad happens, chamomile tea helps you relax. And take a hot bath. That's what I usually do.

upset：イライラして　　chamomile：カモミール（ハーブの一種。鎮静作用が
あるとされる）　　relax：リラックスする　　take a hot bath：熱いお風呂に入る
usually：いつも、普段

mari_tabi

@kathleen_tl Thank you. That seems like a good idea. I'll try the tea idea. I think I still have some in my cabinet.

cabinet：戸棚

A Trip to Canada **Story 1**

3日目

mari_tabi

会社の製品に問題があって、1日中お客様に説明したり謝ったりしてた。もうグッタリ!

mari_tabi

みんな「あなたが責任者だから」って言うの。ええ、そうよ、でもそれのどこがいいの? 誰も私の言うことなんか聞いてくれないんだから!

kathleen_tl

@mari_tabi イライラしないで。何か悪いことがあったら、カモミール・ティーを飲むとリラックスするわよ。それから、あったかいお風呂に入って。私はいつもそうするの。

mari_tabi

@kathleen_tl ありがとう。それはいい考えね。そのお茶を試してみることにする。まだ戸棚にあったと思う。

 mari_tabi Marippe　 kathleen_tl Kathleen Taylor　 patty_coop Patty Cooper　 sweet_pie01 Liu Fei

英語 Twitter 多読術

Story 1 A Trip to Canada

sweet_pie01

We really had a good weekend at Mount Dongbai! I don't think most people have heard of this place but it's really worth visiting!

weekend：週末　Mount Dongbai：ドンバイ山（上海近郊にある山。ハイキングの名所）
most：ほとんどの　　worth -ing：〜する価値がある

---

mari_tabi

@sweet_pie01 Sounds interesting! Hiking is also quite popular among Japanese women. Some women even like to hike alone.

sound 〜：〜のように聞こえる、思える　　popular：人気がある
hike：ハイキングをする　　alone：一人で

---

sweet_pie01

@mari_tabi Exactly the same thing is true in China. We don't care about doing it alone or together with friends. We just do what we want.

exactly：まさに　　true：本当の　　don't care about 〜：〜を気にしない

---

kathleen_tl

Oh, my red roses don't look as good as I expected. Maybe I can replace them with white ones.

as good as 〜：〜ほどよくない　　expect：期待する、望む
replace 〜 with ...：…を〜の代わりにする

A Trip to Canada **Story 1**

sweet_pie01

ドンバイ山で、とってもいい週末を過ごしたの！ この場所を聞いたことのある人は多くないと思うけど、本当に行く価値のある所よ。

mari_tabi

@sweet_pie01 面白そうね！ 日本女性の間でも、ハイキングにとても人気があるの。一人で行く人もいるのよ。

sweet_pie01

@mari_tabi 中国でもまさに同じことが言えるわ。一人か、友達と一緒かなんて気にしない。やりたいことをやるだけよね。

kathleen_tl

あら、うちの赤いバラ、思ったほどきれいじゃないわね。白いやつに替えた方がいいかも。

Marippe

Kathleen Taylor

Patty Cooper

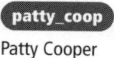

Liu Fei

Story 1 A Trip to Canada

DAY 4

mari_tabi

We've made it! Now there's no problem for me to take my vacation next month. I need to get prepared for the trip to Canada!

> make it：うまくいく、間に合う get prepared for ～：～の準備をする trip：旅行

kathleen_tl

@mari_tabi Good for you! Why don't you fly to Vancouver and then come over to Victoria? It's so close.

> Good for you!：よかったね！　Why don't you ～?：～してはどう？
> fly：飛ぶ、飛行機で行く　come over to ～：～へやってくる　close：近い

kathleen_tl

@mari_tabi You can take a ferry to get to Victoria and there is also a seaplane. It's really worth trying!

> seaplane：水上飛行機（バンクーバーとビクトリアを結ぶホバークラフトのこと）

mari_tabi

OK, I have to find a flight to Vancouver. It's just a one-week vacation, but it'll be a great experience for me!

> experience：経験

sweet_pie01

@mari_tabi Wow, I'm so happy for you. I wish I could have some vacation time too. I'm still so busy and can't take a day off.

> I wish I could ～：～できたらいいのに　vacation time：休み、休暇
> day off：1日休み

A Trip to Canada **Story 1**

4日目

mari_tabi

やった！ これで来月休暇を取るのに何の問題もなくなったわ。カナダ旅行の準備をしなくっちゃ！

---

kathleen_tl

@mari_tabi よかったね！ 飛行機でバンクーバーに行って、それからビクトリアに来れば？ とっても近いわよ。

---

kathleen_tl

@mari_tabi ビクトリアまでフェリーに乗ればいいわ、それから水上飛行機もあるのよ。乗ってみる価値あるわ！

---

mari_tabi

よし、バンクーバー行きのフライトを探さないと。ほんの１週間の休暇だけれど、私とっては素晴らしい経験になりそうね！

---

sweet_pie01

@mari_tabi わあ、よかったね。私も休みが取れたらいいなあ。まだすごく忙しくって、１日休むのだって、できないわ。

Story 1 A Trip to Canada

---

mari_tabi

@sweet_pie01 I think you're working too hard. You really need to find some time to relax. How about some chamomile tea?

---

kathleen_tl

My rosemary is growing OK. It's one of the easiest herbs to take care of. Here's a photo. http://twitpic.com/xxx Take a look at it!

> grow OK：よく育つ、元気に育つ　　here's 〜：これが〜、こちらに〜（ツイッターの写真サイトにリンクしている）　　take a look at 〜：〜を見てみる

---

kathleen_tl

Oh, I got a message from my boyfriend in Toronto. He'll be coming back to Victoria as usual this weekend!

> as usual：いつものように

DAY 5

mari_tabi

I'm going to make reservations for hotels on the Internet. I found a pretty one but it's a bit over my budget. What should I do?

> make a reservation：予約する（reservationは「予約」）　　pretty：きれいな
> a bit：少し、ちょっと　　over one's budget：予算オーバー（budgetは「予算」）

---

kathleen_tl

@mari_tabi I think hotels with an Inner Harbour view would be great. And everything is around the harbour. Quite convenient!

> Inner Harbour：インナー・ハーバー（ビクトリアの街にある内湾のこと。innerは「内側の」、harbourは「港湾」）　　view：眺め、景観

A Trip to Canada **Story 1**

mari_tabi

@sweet_pie01 あなた、働きすぎなのよ。本当にリラックスする時間を見つけた方がいいわ。カモミール・ティーなんてどう？

kathleen_tl

ローズマリーはよく育ってる。育てるのがすごく簡単なハーブなのよね。これが写真。 http://twitpic.com/xxx 見てね！

kathleen_tl

あっ、トロントの彼からメッセージが来た。今週末も、いつものようにビクトリアに帰ってくるって！

5日目

mari_tabi

インターネットでホテルを予約しよう。きれいなのを見つけたんだけれど、ちょっと予算をオーバーしてるのよね。どうしよう？

kathleen_tl

@mari_tabi インナー・ハーバーが見えるホテルがいいと思うわ。ハーバーの周りにいろんな物があるし。とっても便利よ！

 mari_tabi Marippe   **kathleen_tl** Kathleen Taylor   **patty_coop** Patty Cooper   **sweet_pie01** Liu Fei

Story 1 A Trip to Canada

mari_tabi

I got my plane tickets! I found a really cheap site on the Internet and it was quite easy to use. Actually it only took a few minutes!

plane ticket：飛行機のチケット　easy to use：使うのが簡単な
a few minutes：数分

kathleen_tl

@mari_tabi When do you arrive in Victoria? I think I can show you around. How about going to Butchart Gardens?

mari_tabi

@kathleen_tl Thanks! I'll let you know later. By the way, I saw the pic of your rosemary. I have nothing to tell you about my garden yet.

let ～ know：～に知らせる　by the way：ところで　pic=picture：写真

kathleen_tl

Tonight I'm going to cook some chicken and if I add some rosemary, it'll taste great. It's good to have your own herb garden!

cook：料理する　add：加える　taste great：おいしい　own：自分自身の

sweet_pie01

My friend J was late again today. Amy says it's nothing but I really think we should do something about it.

late：遅れて　nothing：何でもない

A Trip to Canada **Story 1**

> **mari_tabi**
>
> 飛行機のチケット取れた！ インターネットですごく安いサイトを見つけたの、使い方はとっても簡単だったわ。実際、数分しかかからなかった！

> **kathleen_tl**
>
> **@mari_tabi** いつビクトリアに着くの？ 案内してあげられると思うわ。ブッチャート・ガーデンに行くのはどう？

> **mari_tabi**
>
> **@kathleen_tl** ありがとう！ 後で教えるね。ところで、あなたのローズマリーの写真見たわ。私の庭については、まだ何も話せることがないの。

> **kathleen_tl**
>
> 今晩、チキンを料理するけれど、ローズマリーを加えたら、おいしくなるわね。自分のハーブ・ガーデンがあるっていいものね！

> **sweet_pie01**
>
> 友達のJが、今日また遅れてきたの。エイミーは何でもないって言うけれど、私は本当に何とかすべきだと思うわ。

mari_tabi
Marippe

kathleen_tl
Kathleen Taylor

patty_coop
Patty Cooper

sweet_pie01
Liu Fei

Story 1 A Trip to Canada

mari_tabi

@sweet_pie01 Don't get so upset! I think Amy is right. I know people who are always late, but all we can do about it is just wait.

> right：正しい　　always：いつも　　just wait：待つだけ

DAY 6

mari_tabi

Today I'm going shopping in Shibuya. I need to get something warm for my trip to Canada.

> go shopping：買い物に行く　　something warm：何か暖かい物　　trip：旅行

mari_tabi

Oops, I bought another wool scarf. What should I do with all these scarves? Maybe my sis wants some of them.

> Oops!：おっと！　　wool scarf：マフラー、ウールのスカーフ
> scarves：scarfの複数形　　sis＝sister：姉、妹

kathleen_tl

Got the phone call from Tom. He's at Vancouver airport now and I'm getting ready to pick him up. What does he want to eat tonight?

> pick 〜 up：（車で）迎えに行く

A Trip to Canada **Story 1**

mari_tabi

@sweet_pie01 そんなに怒らないで！ エイミーが正しいと思うな。いつも遅れる人を知っているけれど、私たちにできるのは、待つことだけなのよ。

6日目

mari_tabi

今日は渋谷に買い物に行くの。カナダ旅行のために、何か暖かい物を買っておかないと。

mari_tabi

おーっと、またマフラー買っちゃった。こんなにマフラーがあってどうしよう？ 妹がいくつか欲しいかもね。

---

kathleen_tl

トムから電話だった。今バンクーバー空港にいるから、迎えに行く準備をしてるところ。彼、今晩何を食べたいかな？

 **mari_tabi** Marippe  **kathleen_tl** Kathleen Taylor  **patty_coop** Patty Cooper   **sweet_pie01** Liu Fei

Story 1 A Trip to Canada

mari_tabi

@kathleen_tl I've never had a "long-distance relationship." I guess it's not that easy, but seeing someone every weekend sounds romantic.

> long-distance relationship：遠距離恋愛（distanceは「距離」、relationshipは「関係」）
> guess ～：～と思う　　someone：誰か　　every weekend：毎週末
> romantic：ロマンチックな

kathleen_tl

What should I wear? I wore my green dress last week, so my white blouse would be better. Oh, it's still in the laundry!

> dress：ワンピース　　blouse：ブラウス　　laundry：クリーニング店

kathleen_tl

After the movie, we're going to Tony's for dinner. I should make a reservation? They are always busy on Saturday night.

> Tony's：トニーの店　　busy：（店が）混んでいる

sweet_pie01

@mari_tabi Are you bringing pocket warmers with you? I've heard they're originally from Japan but now we can buy them here too.

> pocket warmer：ポケットカイロ　　originally：元々

mari_tabi

@sweet_pie01 I thought pocket warmers were common all over the world. So I should bring some, I may not be able to find them in Canada.

> common：一般的な、共通の　　all over the world：世界中で　　bring：持って行く
> be able to ～：～できる

A Trip to Canada **Story 1**

> **mari_tabi**
>
> **@kathleen_tl** 私、「遠距離恋愛」ってしたことない。そんなに簡単じゃないんだろうけれど、毎週末恋人に会うのって、ロマンチックね。

> **kathleen_tl**
>
> 何を着て行こう？ 先週はグリーンのワンピースを着たから、白いブラウスの方がいいかな。あっ、まだクリーニングに出してる！

> **kathleen_tl**
>
> 映画の後、トニーの店に夕食に行くの。予約した方がいいかな？ 土曜の晩はいつも混んでいるから。

> **sweet_pie01**
>
> **@mari_tabi** ポケットカイロは持っていくの？ 元々日本から来たって聞いたんだけど、今はここでも買えるんだよ。

> **mari_tabi**
>
> **@sweet_pie01** ポケットカイロって、世界中でよくあるんだと思ってた。じゃあいくつか持っていこう、カナダで見つからないかもしれないから。

 Marippe　 Kathleen Taylor　 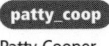 Patty Cooper　 Liu Fei

Story 1 A Trip to Canada

DAY 7

mari_tabi

It's my day off today! I'm done with my cleaning and the laundry. Well, what should I do next?

> be done with ～：～を終えている　　cleaning：掃除　　laundry：洗濯

mari_tabi

Lately, it's getting more difficult to ask friends to go out on weekends. They usually spend their day off with their husband or children.

> lately：最近　　get＋（比較級）：より～する　　go out：外出する　　spend：過ごす
> husband：夫

mari_tabi

I don't think marriage is so important but it'd be great if you have a family and can enjoy time together.

> marriage：結婚　　it'd be great if ～：～だったらすごい

kathleen_tl

Tom is leaving today. I think I'm going to see him off at the airport. Whew, how long will this continue?

> leave：去る　　see＋（人）＋off：（人）の見送りをする　　whew：ふう、やれやれ
> how long：どれくらいの期間　　continue：続く

kathleen_tl

I really shouldn't complain about this. Toronto is just 4 and a half hours away by air, and it's still in the same country.

> complain：文句を言う　　by air：飛行機で

7日目

mari_tabi

今日はお休み！ 掃除と洗濯は終わった。さて、次は何をしよう？

---

mari_tabi

最近、週末に友達を外出に誘うのが難しくなってきてるんだよね。みんな普段、だんなさんや子どもと一緒に過ごすから。

---

mari_tabi

結婚がそんなに大事だとは思わないけれど、家族がいて、一緒に楽しい時間を過ごせたらいいだろうな。

---

kathleen_tl

トムは今日行ってしまう。空港に見送りに行くことにしよう。ああ、こんなこといったいどれくらい続くんだろう？

---

kathleen_tl

本当は文句を言うべきじゃないわね。トロントは飛行機で4時間半離れているだけだし、まだ同じ国の中なんだし。

mari_tabi
Marippe

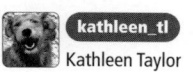
kathleen_tl
Kathleen Taylor

patty_coop
Patty Cooper

sweet_pie01
Liu Fei

Story 1 A Trip to Canada

mari_tabi

@kathleen_tl Does he really come back every weekend? If so, I think you are very lucky. It's not everyone who can do that.

if so：もしそうなら　　lucky：ラッキーな

kathleen_tl

@mari_tabi Yeah, I guess so. And he brings something back for me from Toronto every week!

bring ～ back：～を持ち帰る

mari_tabi

@kathleen_tl Sounds like you have a great boyfriend! I hope I have a chance to meet him while I'm in Canada.

chance：機会、チャンス　　while ～：～の間に

DAY 8

mari_tabi

Whew, that was close! I almost missed the train. My alarm clock didn't go off or I don't remember turning it off.

close：近い、きわどい　　miss the train：電車に乗り損ねる
alarm clock：目覚まし時計　　go off：（時計が）鳴る

mari_tabi

I made it to the office. It's good nobody noticed that I'm later than usual. Or was the boss watching me when I got in?

make it to ～：～に間に合う　　notice：気付く　　usual：いつも、普段
or：それとも、もしくは　　boss：上司　　get in ～：～に入る

mari_tabi

@kathleen_tl 彼は本当に毎週末帰ってくるの？ もしそうなら、あなたすごいラッキーね。誰でもできることじゃないわよ。

kathleen_tl

@mari_tabi ええ、そう思う。それに、毎週トロントから私に何か持って帰って来てくれるの！

mari_tabi

@kathleen_tl ステキなボーイフレンドみたいね！ カナダにいる間に、その人に会う機会があるといいな。

8日目

mari_tabi

ふう、やばかった！ 電車に乗りそこねるところだった。目ざまし時計が鳴らなかったか、止めたの覚えてないのね。

mari_tabi

会社に間に合った。よかった、いつもより遅いのに、誰も気付いてない。それか、入ったとき上司が見てた？

mari_tabi
Marippe

kathleen_tl
Kathleen Taylor

patty_coop
Patty Cooper

sweet_pie01
Liu Fei

Story 1　A Trip to Canada

sweet_pie01

I went to a big shopping mall which opened recently. It's huge! Please check the pic. http://twitpic.com/xxx

recently：最近　　huge：巨大な

mari_tabi

@sweet_pie01 I saw your photo! Looks so exciting, I'd love to go there sometime. And I noticed Starbucks is everywhere around the world.

exciting：面白い、ワクワクする　　sometime：いつか　　everywhere：どこにでも
around the world：世界中

kathleen_tl

Tom is now back in Toronto. It's been a while since the last time I went there. Maybe I should go over sometime.

back：戻って、帰って　　It's been a while since 〜：〜からしばらく経っている（it's =it has）　　last time：最後に、前回　　go over 〜：〜に行く

kathleen_tl

The flowers in my garden are growing too fast. Is it because it's too warm this year? I wonder if I need to do something about it.

too fast：早すぎる

mari_tabi

@kathleen_tl I'm having the same problem here. I heard you need to water a little more when it's warmer. I'll try that.

water：水をやる（動詞）

kathleen_tl

@mari_tabi Hey, I have some friends who are interested in gardening. Maybe we can go to Butchart Gardens together.

sweet_pie01
最近オープンした新しいショッピングモールに行ってきたの。巨大！写真見てね。http://twitpic.com/xxx

mari_tabi
@sweet_pie01 写真見たよ！ 面白そうだね、いつか行ってみたいな。それと、スターバックスって世界中どこにでもあるんだね。

kathleen_tl
トムがトロントに帰ってる。私が最後にトロントに行ってから、しばらく経ってるな。ときどき私が行った方がいいのかも。

kathleen_tl
庭の花がずいぶん早く育ってる。今年は暖かすぎるから？ 何かした方がいいのかしら。

mari_tabi
@kathleen_tl こちらでもおんなじ。普段より暖かいときは、少し多めに水をやるといいって聞いたわ。私はそうしてみる。

kathleen_tl
 @mari_tabi ねえ、友達で、ガーデニングに興味がある人がいるの。一緒にブッチャート・ガーデンに行きましょうか。

mari_tabi
Marippe

kathleen_tl
Kathleen Taylor

patty_coop
Patty Cooper

sweet_pie01
Liu Fei

Story 1   A Trip to Canada

DAY 9

mari_tabi

In Vancouver, I'd love to go to Granville Island for shopping. I heard there are some good restaurants there too.

Granville Island：グランビル島（市場やレストラン、お店等が集まったエリア）

kathleen_tl

@mari_tabi In Vancouver, why don't you visit Gastown? It's the oldest part of Vancouver and you can see the steam clock on the street!

Gastown：ガスタウン（「バンクーバー発祥の地」とされる）
steam clock：蒸気時計（蒸気で時を知らせる）

mari_tabi

And in Victoria, I'm interested in Craigdarroch Castle. It was built in the 19th century, by a wealthy immigrant, right?

Craigdarroch Castle：クレイダーロック城（ビクトリアの中心部にある屋敷。現在は博物館）   wealthy：裕福な   immigrant：移民   〜, right?：〜でしょ？

kathleen_tl

@mari_tabi That's right! It's Victorian style and the stained-glass windows are really beautiful.

sweet_pie01

Listen to this! I was promoted to district manager. I thought Jeremy would be the next manager, so I'm really surprised and happy!

be promoted to 〜：〜に昇進する   district：地域の

A Trip to Canada **Story 1**

9日目

mari_tabi

バンクーバーでは、グランビル島に買い物に行きたいな。いいレストランもあるって聞いたし。

kathleen_tl

@mari_tabi バンクーバーでは、ガスタウンに行ってみれば？ バンクーバーの一番古い場所で、通りで蒸気時計が見られるの！

mari_tabi

それにビクトリアでは、クレイダーロック城に興味があるの。19世紀に、裕福な移民が建てたんでしょ？

kathleen_tl

@mari_tabi そうなの！ ビクトリア朝様式でステンドグラスの窓が本当にきれい。

sweet_pie01

これ聞いて！ 地域マネジャーに昇進したの！ ジェレミーが次のマネジャーになると思ってから、本当に驚いたし、うれしい！

| mari_tabi
Marippe | kathleen_tl
Kathleen Taylor | patty_coop
Patty Cooper | sweet_pie01
Liu Fei |

英語Twitter多読術 **35**

Story 1 A Trip to Canada

mari_tabi

@sweet_pie01 Congratulations! You've been working so hard and I think you deserve it. Good for you!

> Congratulations!：おめでとう！　deserve 〜：〜に値する

patty_coop

@mari_tabi Hello! I'm a friend of Kathleen's and I heard you're interested in gardening. I have my own garden too.

mari_tabi

@patty_coop Thank you for following me! It's nice to know someone new through Twitter. My small balcony garden looks fine today!

> thank you for 〜：〜をありがとう　follow：（ツイッターで）フォローする
> look 〜：〜のように見える

DAY 10

mari_tabi

Big news! One of my colleagues will get married and leave the company next week. It's so sudden, nobody knew about it except the boss!

> colleague：同僚　get married：結婚する　leave the company：会社を辞める
> sudden：突然の　except 〜：〜以外に

A Trip to Canada **Story 1**

> **mari_tabi**
>
> **@sweet_pie01** おめでとう！ 一生懸命働いてきたから、当然のごほうびね。よかったね！

> **patty_coop**
>
> **@mari_tabi** こんにちは！ 私、キャスリーンの友達で、あなたがガーデニングに興味があるって聞いたの。私も自分の庭を持ってるのよ。

> **mari_tabi**
>
> **@patty_coop** フォローしてくれてありがとう！ ツイッターで新しい人と知り合うのっていいわね。私の小さなバルコニーの庭は、今日は調子よさそう！

10日目

> **mari_tabi**
>
> ビッグニュース！ 同僚の一人が結婚して、来週会社を辞めるんだって。すごい突然、上司以外は誰も知らなかったのよ！

| **mari_tabi** | **kathleen_tl** | **patty_coop** | **sweet_pie01** |
| Marippe | Kathleen Taylor | Patty Cooper | Liu Fei |

Story 1 A Trip to Canada

> **mari_tabi**
>
> Maybe we should throw a farewell party and give her a gift. This is unexpected. I hope it doesn't cost a lot.
>
> throw：（パーティーを）開く　　farewell party：さよならパーティー
> unexpected：予期しない　　cost a lot：お金がたくさんかかる

> **kathleen_tl**
>
> I'm waiting for Patty to go out for dinner. I've been here for 15 minutes and I hope she gets here soon.
>
> go out for dinner：夕食に出る　　get here：ここに着く

> **kathleen_tl**
>
> It's kind of a party and I wish Tom could be here. I always have to ask Patty instead. Oh, here she comes!
>
> I wish ～ could ...：～が…できるといいのに　　instead：代わりに
> here ～ come：ほら、～が来た

> **mari_tabi**
>
> **@kathleen_tl** You usually go to a party with your boyfriend? Isn't it common to go with your female friends?
>
> common：一般的な、共通の　　female：女性の、女性

> **kathleen_tl**
>
> **@mari_tabi** Nothing wrong with going out with your friends but I think it's more relaxing when you're with your husband or boyfriend.
>
> nothing wrong with ～：～については何も問題ない　　relaxing：落ち着く、ほっとする

A Trip to Canada **Story 1**

> **mari_tabi**
>
> さよならパーティーを開いてプレゼントをあげるといいのかな。これは想定外だわ。あんまりお金かからないといいけど。

> **kathleen_tl**
>
> 夕食に行くのにパティを待っているところ。もうここに15分いる。すぐに来てくれるといいけど。

> **kathleen_tl**
>
> 一種のパーティーだから、トムがいてくれればよかったのに。いつも代わりにパティに頼まないといけない。あ、来た！

> **mari_tabi**
>
> **@kathleen_tl** パーティーにはいつもボーイフレンドと行くの？ 女の友達と行くのは一般的じゃないの？

> **kathleen_tl**
>
> **@mari_tabi** 友達と行くのは全然かまわないんだけれど、夫かボーイフレンドと一緒の方が落ち着くわね。

mari_tabi	kathleen_tl	patty_coop	sweet_pie01
Marippe	Kathleen Taylor	Patty Cooper	Liu Fei

Story 1 A Trip to Canada

mari_tabi

@kathleen_tl Interesting! In Japan, nobody cares who you go to a party with. But of course it's more fun if you're with your boyfriend!

care：気にする　　who you go to a party with：誰とパーティーに行くか
of course：もちろん　　fun：楽しい、楽しみ

kathleen_tl

Actually I am not as social as before, since Tom went to Toronto. We used to have house parties all the time!

social：社交的な　　not as 〜 as before：前ほど〜でない

DAY 11

mari_tabi

I need to buy a few more things for my trip to Canada. I want some warmer socks, a shirt, and maybe I should buy a new hat!

a few more：後いくつかもっと　　socks：靴下

mari_tabi

@sweet_pie01 I'm going to bring a gift to someone in Canada. If it was you, what would you like from Japan?

gift：贈り物、プレゼント　　if it was you：もしあなただったら

sweet_pie01

@mari_tabi Cosmetics, of course! But if it's for someone in Canada, I think a Japanese handkerchief or fan would be nice.

cosmetics：化粧品　　Japanese handkerchief：手ぬぐい　　fan：扇子

mari_tabi
@kathleen_tl 面白いね！ 日本では、誰と一緒にパーティーに行くかなんて、誰も気にしないわよ。でももちろん、ボーイフレンドと一緒の方が楽しいよね！

kathleen_tl
実はトムがトロントに行ってから、あまり人と会ってないの。前はしょっちゅうホームパーティーを開いていたのに！

11日目

mari_tabi
カナダに旅行するのに、後いくつか買い物をしないと。暖かい靴下と、シャツと、新しい帽子も買おうかな！

mari_tabi
@sweet_pie01 カナダの人にお土産を持っていくの。あなただったら、日本からのお土産は何がいい？

sweet_pie01
@mari_tabi 化粧品、もちろん！ でも、カナダの人なら、手ぬぐいや扇子がいいんじゃない。

---

mari_tabi Marippe | **kathleen_tl** Kathleen Taylor | **patty_coop** Patty Cooper | **sweet_pie01** Liu Fei

Story 1 A Trip to Canada

kathleen_tl

What do I know about Japan? Sushi, tempura, kimono ... and famous cherry blossoms! I saw them in a photo before.

cherry blossom：桜（blossomは「花」）

kathleen_tl

I heard Japanese people love nature so much and they make very unique gardens. I'd love to see them sometime.

unique：独特の

mari_tabi

@kathleen_tl I've visited some Japanese gardens before. I think I can show you some photos when I come to Canada.

kathleen_tl

@mari_tabi Thanks! I heard some gardens are made of only rocks and stones, without flowers or plants. I'm looking forward to seeing them!

rock：岩　　look forward to -ing：〜することを楽しみにする

patty_coop

@mari_tabi The only Japanese gardens I know are in Victoria and San Francisco. I'd love to see real ones in Japan sometime.

real one：本物

A Trip to Canada **Story 1**

kathleen_tl

私、日本について何を知ってるかしら？ 寿司、天ぷら、着物……それに有名な桜！ 前に写真で見たわ。

kathleen_tl

日本の人は自然をとても愛していて、とても独特の庭を造るそうね。いつか見てみたいわ。

mari_tabi

@kathleen_tl 前に、日本庭園に行ったことがあるわよ。カナダに行ったら、写真を見せるわね。

kathleen_tl

@mari_tabi ありがとう！ 花や緑がなくて、岩と石だけで造られている庭があるんですってね。見るのが楽しみ！

patty_coop

@mari_tabi 私が知っている日本庭園は、ビクトリアとサンフランシスコにあるのだけ。いつか日本で本物を見てみたいな。

---

mari_tabi Marippe
kathleen_tl Kathleen Taylor
patty_coop Patty Cooper
sweet_pie01 Liu Fei

Story 1　A Trip to Canada

DAY 12

mari_tabi

I'm going to an English lesson today. It's been a while. I wonder if I still remember how to speak.

> how to speak：話し方、どうやって話すか

mari_tabi

Whew, what a tough lesson! I could hardly speak. I think I really need to practice English more before I leave for Canada.

> practice：練習する　　leave for 〜：〜に向けて出発する

kathleen_tl

@mari_tabi That reminds me of my French lessons. It's our country's other official language, but I've never been good at it.

> remind 〜 of ...：〜に…を思い出させる　　other：別の
> official language：公用語　　be good at 〜：〜が得意な、上手な

patty_coop

@mari_tabi Canada is a mixed country, so you don't have to worry about it. You might even hear Chinese more often than English!

> mixed country：多民族の国（mixedは「混ざった」）　　might 〜：〜かもしれない
> even 〜：〜さえ（ある）　　often：しばしば

kathleen_tl

Oh, I need to buy some more bread at the supermarket since Tom will be back in Victoria soon.

> since：なぜなら

A Trip to Canada **Story 1**

12日目

mari_tabi

今日は英語のレッスンに行くの。しばらくぶりだな。どうやってしゃべるか、まだ覚えてるかな。

mari_tabi

ふう、レッスン、キツかった！ ほとんどしゃべれなかった。カナダに行く前に、もっと本気で英語の練習をしないといけないな。

kathleen_tl

@mari_tabi 自分のフランス語のレッスンのこと思い出した。フランス語は私たちの国のもう一つの公用語だけれど、私は上手だったためしがないの。

patty_coop

@mari_tabi カナダは多民族の国だから、あんまり心配する必要ないわよ。ここでは英語より中国語の方をよく耳にするかもよ！

kathleen_tl

ああ、スーパーでもっとパンを買っておかないと、トムがすぐビクトリアに帰ってくるから。

---

mari_tabi Marippe   **kathleen_tl** Kathleen Taylor   **patty_coop** Patty Cooper   **sweet_pie01** Liu Fei

Story 1 A Trip to Canada

kathleen_tl

I forgot which he likes better, rye bread or raisin bread? Maybe I should buy both of them.

rye：ライ麦の　both：両方

sweet_pie01

Oh, this is awful. I have a runny nose and a terrible cough. I think I've got a cold. Maybe I should stay home this weekend.

awful：ひどい、大変な　have a runny nose：鼻水が止まらない　terrible：ひどい
cough：せき　I've got a cold.：風邪をひいた（I've got＝I have）

mari_tabi

@sweet_pie01 Are you OK? Staying home all weekend would be boring but you could catch up on the movies you've been missing. Get well soon!

boring：退屈な　catch up on ～：～に追いつく　miss：見逃す
get well：回復する

DAY 13

mari_tabi

I'm going to ask T-san to cover me during my vacation. Oh, no! I just got something I have to do myself. I wonder if I can finish it.

during ～：～の間

mari_tabi

I have to work overtime again. My flight is the day after tomorrow, and I haven't even packed yet!

work overtime：残業する　the day after tomorrow：あさって　pack：荷造りする

A Trip to Canada **Story 1**

kathleen_tl

彼はどっちが好きか忘れちゃった。ライ麦パンかレーズンパンか？ 両方買っておいた方がいいかな。

sweet_pie01

ああ、これはヒドイ。鼻水が止まらなくて、せきがヒドイ。風邪をひいたみたい。今週の週末は家にいた方がいいかな。

mari_tabi

@sweet_pie01 大丈夫？ 週末ずっと家にいるのは退屈だろうけど、見逃した映画に追いつけるんじゃない。早くよくなってね！

13日目

mari_tabi

休みの間中、Tさんに代わりをやってくれるよう頼もう。あっ、しまった！ 自分でやらないといけないことがあった。終わらせられるかなあ。

mari_tabi

また残業しないといけない。飛行機はあさってなのに、まだ荷造りもしていない！

mari_tabi Marippe	**kathleen_tl** Kathleen Taylor	**patty_coop** Patty Cooper	**sweet_pie01** Liu Fei

Story 1 A Trip to Canada

sweet_pie01

@mari_tabi Hey, seems like you're busy again. These things usually happen before a vacation, but you'll manage it somehow.

happen：起こる　manage：何とかする　somehow：どうにかして

mari_tabi

@sweet_pie01 I hope so! Oops, no time for Twitter. I have to get back to work. Only 2 hours until the last train!

I hope so.：そうだといいな　last train：終電

kathleen_tl

What?! I can't believe it. Tom is not coming back and he only sent me a few words about it. I don't know what to say.

a few words：2、3言　I don't know what to say.：何と言えばいいか分からない

mari_tabi

@kathleen_tl Too bad! I think your boyfriend must be busy with his job. That kind of thing happens quite often in Japan.

too bad：残念

kathleen_tl

I've already prepared so much food. I guess I'll invite some friends over for dinner Saturday night? I'll grill some chicken.

already：すでに　guess 〜：〜と思う　invite 〜 over ...：〜を…に招く
grill：グリルで焼く

patty_coop

I heard the weather will be nice this weekend, so it should be good for gardening. And I'll stop by Kathleen's!

A Trip to Canada **Story 1**

sweet_pie01

@mari_tabi また忙しくなったみたいね。休みの前はたいていそういうことがあるけれど、どうにかなるもんよ。

mari_tabi

@sweet_pie01 そうだといいけど！ おっと、ツイッターやってる場合じゃない。仕事に戻らないと。終電まで後２時間しかない！

kathleen_tl

何ですって?!　信じられない。トムが戻ってこなくて、そのことについて簡単に言ってきただけ。もう何て言ったらいいか分からないわ。

mari_tabi

@kathleen_tl 残念！　きっとお仕事忙しいんでしょうね。日本ではそういうことよくあるのよ。

kathleen_tl

もう食べ物たくさん用意しちゃったのに。土曜の夜は友達を夕食に呼ぼうかしら？　チキンを焼くつもりなの。

patty_coop

週末は天気がいいっていうから、きっとガーデニング日和だわね。それから、キャスリーンの所に寄ろうっと！

---

mari_tabi	kathleen_tl	patty_coop	sweet_pie01
Marippe	Kathleen Taylor	Patty Cooper	Liu Fei

英語Twitter多読術

Story 1 A Trip to Canada

DAY 14

mari_tabi

OK. I'm going to leave the office at 5 pm today to get ready for tomorrow's departure. Wow, this is so exciting!

get ready for 〜：〜の準備をする　　departure：出発

mari_tabi

@kathleen_tl Hi, I'm leaving for Canada soon. I'll arrive in Vancouver tomorrow as scheduled. Looking forward to seeing you!

arrive in 〜：〜に到着する　　as scheduled：スケジュールどおりに

kathleen_tl

@mari_tabi All right! Please tweet me when you arrive in Vancouver. I'll come with Patty to see you in Victoria!

patty_coop

@mari_tabi So you'll be arriving on Harbour Air, right? It should be very exciting. I hope you enjoy the flight.

Harbour Air：ハーバー・エア（バンクーバーとビクトリアを結ぶ水上飛行機の運行会社、もしくはその飛行機のこと）

mari_tabi

@kathleen_tl @patty_coop Thank you! I'm really looking forward to seeing you guys. Hope you have a good weekend!

you guys：あなたたち（口語的な言い方。guyは単数で「あいつ」と男性を指す）

14日目

mari_tabi

よし、今日は5時に会社を出て、明日の出発の準備をしよう。わあ、ワクワクしちゃう！

mari_tabi

@kathleen_tl ハイ！ もうすぐカナダに出発します。明日予定通りバンクーバーに着きます。会えるのを楽しみにしているね！

kathleen_tl

@mari_tabi 了解！ バンクーバーに着いたらツイートしてね。ビクトリアで、パティと一緒に会いに行くから！

patty_coop

@mari_tabi じゃあ、「ハーバー・エア」で着くのね？ きっと面白いわよ。フライトを楽しんでね。

mari_tabi

@kathleen_tl @patty_coop ありがとう！ 二人に会えるのが本当に楽しみ。いい週末をね！

---

mari_tabi Marippe　　**kathleen_tl** Kathleen Taylor　　**patty_coop** Patty Cooper　　**sweet_pie01** Liu Fei

Story 1 A Trip to Canada

kathleen_tl

It's weird. I should've been really busy this weekend, but suddenly, there's nothing to do. Maybe I'll go to a flower market.

| weird：奇妙な、変な　　suddenly：突然　　market：市場 |

mari_tabi

Oh, I think I forgot to buy toothpaste for my trip. Is there anything else? I'm going to buy everything on my way home.

| toothpaste：歯磨き粉　　else：ほかに　　on my way home：帰宅途中に |

mari_tabi

Whew, I made it. Everything is all set. Now, the only thing I have to do tomorrow morning is go to the airport!

| all set：準備万端 |

DAY 15

mari_tabi

I'm in Vancouver now! It's weird, still the same day that I left Tokyo, and it's still bright outside.

| bright：明るい　　outside：外は |

mari_tabi

@kathleen_tl Hi! I just arrived at the hotel in Vancouver. The city is much bigger than I thought. So many tall buildings!

kathleen_tl

@mari_tabi Welcome to Canada! How was the flight? I think you'll have nice weather in Vancouver tomorrow.

kathleen_tl

変な感じ。今週末は本当に忙しいはずだったんだけれど、突然、やることがなくなっちゃった。花の市場にでも行こうかな。

mari_tabi

あっ、旅行用の歯磨き粉を買うの忘れちゃった。ほかに何かないかな？家に帰る途中に、みんな買っていこう。

mari_tabi

ふう、間に合った。すべて準備万端。さて、明日の朝やらなければならないのは、空港に行くことだけだわ！

15日目

mari_tabi

バンクーバーに着いた！ 変な感じ、東京を出たのと同じ日なのに、まだ外が明るいんだもの。

mari_tabi

@kathleen_tl ハイ！ バンクーバーのホテルに着いたところ。思ってたよりずっと大きい街ね。大きなビルがいっぱい！

kathleen_tl

@mari_tabi カナダへようこそ！ 空の旅はどうだった？ 明日のバンクーバーは天気がいいみたいよ。

---

mari_tabi	kathleen_tl	patty_coop	sweet_pie01
Marippe	Kathleen Taylor	Patty Cooper	Liu Fei

Story 1 A Trip to Canada

patty_coop

@mari_tabi If you go a little outside of the downtown area, you'll be surrounded by nature. That's why people like Vancouver.

outside of 〜：〜の外　　be surrounded by 〜：〜に囲まれて

sweet_pie01

@mari_tabi Hey, are you in Canada now? Should be colder than Tokyo, right? I hope you have a great time there!

mari_tabi

@sweet_pie01 Thanks. It's kind of warm here. I heard Vancouver is a safe city, so tomorrow I think I'll walk around town by myself.

by myself：独りで、自分で

kathleen_tl

After the rain yesterday, my garden seems really nice. Now I may be able to show it to Mari, if she stops by.

patty_coop

I have to let my husband know that I'm using the car the day after tomorrow, since a friend of mine from Japan will be here!

let 〜 know ...：〜に…を知らせる

A Trip to Canada **Story 1**

patty_coop
@mari_tabi 街の中心部からちょっと離れると、自然に囲まれるの。そこがバンクーバーのいい所なのよ。

sweet_pie01
@mari_tabi ねえ、今カナダなの？ 東京より寒いんでしょ？ 楽しめるといいね！

mari_tabi
@sweet_pie01 ありがとう。ここはなんだかあったかいわよ。バンクーバーは安全な街だって聞いたから、明日は一人で街を歩いてみるつもり。

kathleen_tl
昨日雨が降ったから、今日はお庭が元気に見えるわ。マリが寄ってくれたら見せられるかもね。

patty_coop
だんなに言っておかないと。日本から友達が来るから、あさっては車使うって！

	mari_tabi		kathleen_tl		patty_coop		sweet_pie01
	Marippe		Kathleen Taylor		Patty Cooper		Liu Fei

Story 1 A Trip to Canada

DAY 16

mari_tabi

Wow, I saw so many Chinese people in town, but maybe some of them were Japanese. I see why people say this is an immigrant city.

immigrant city：移民の街（immigrantは「移民」）

kathleen_tl

@mari_tabi I have many friends in Vancouver, some of them are originally from China or India. You can eat so many kinds of food there!

originally：元々

sweet_pie01

@mari_tabi That reminds me that one of my aunts went to Vancouver and married a Canadian guy. Maybe I can see her if I go there.

aunt：叔母（伯母）　　marry＋(人)：(人)と結婚する　　guy：男性、やつ

mari_tabi

I'm at the steam clock in Gastown. Please check the pic! http://twitpic.com/xxx And near this area is a big Chinatown.

kathleen_tl

@mari_tabi I saw the pic! The clock is kind of funny, isn't it? Some people say that it looks like it's pissed off.

funny：おかしい、笑える　　pissed off：カンカンに怒って

A Trip to Canada **Story 1**

16日目

mari_tabi

わあ、街で中国系の人をたくさん見かけたけど、多分中には日本人もいるんだわ。なんでここが移民の街って呼ばれるか分かったわ。

---

kathleen_tl

@mari_tabi バンクーバーには友達がたくさんいて、中には元々中国やインドから来た人もいるの。バンクーバーでは、すごくいろんな種類の食べ物が食べられるのよ！

---

sweet_pie01

@mari_tabi そういえば叔母の一人がバンクーバーに行って、カナダ人と結婚したのを思い出した。バンクーバーに行ったら会えるかな。

---

mari_tabi

ガスタウンの蒸気時計の所にいます。写真見てね！
http://twitpic.com/xxx この近くに大きなチャイナタウンがあるの。

---

kathleen_tl

@mari_tabi 写真見たわよ！ あの時計、なんだか笑えるわよね？ 時計が怒ってるみたいだっていう人もいるの。

| **mari_tabi** Marippe | **kathleen_tl** Kathleen Taylor | **patty_coop** Patty Cooper | **sweet_pie01** Liu Fei |

Story 1 A Trip to Canada

patty_coop

@mari_tabi So you're in Vancouver now. I hope you can visit Queen Elizabeth Park, but one day would not be enough to see it all.

> Queen Elizabeth Park：クイーン・エリザベス公園（バンクーバーの名所の一つ）
> to see it all ：いろいろな物を見る

mari_tabi

I'm at Grandville Island. A great place for shopping. And they have a beer factory too! I think I'm going to try a Canadian beer.

> beer factory：ビール工場

kathleen_tl

Well, I haven't been to Vancouver in a while. Maybe I can go there with Tom next time he comes back.

> in a while：しばらくの間

DAY 17

mari_tabi

I'm taking a seaplane to get to Victoria. Wow, what a view! It's just like a scene in a movie! Here's the pic. http://twitpic.com/xxx

> take a seaplane：水上飛行機に乗る　　wow：わあ　　what a view：すごい眺め
> scene：シーン、光景　　Here's 〜：これが〜、ほら〜

sweet_pie01

@mari_tabi Amazing! I want to take that seaplane too! Please let me know what it's like. Isn't it scary?

> amazing：すごい、素晴らしい　　what it's like：どんな感じか　　scary：怖い

A Trip to Canada **Story 1**

patty_coop

@mari_tabi じゃあ、今はバンクーバーなのね。クイーン・エリザベス公園に行けるといいんだけど、1日じゃ全部見るにはとても足りないわよね。

mari_tabi

グランビル島にいます。買い物するのにサイコーの所。ビール工場もあるの！ カナダビールを試してみようかな。

kathleen_tl

そうね、バンクーバーにはしばらく行ってないわ。今度トムが帰ってきたときに、一緒に行ってもいいわね。

17日目

mari_tabi

ビクトリアへ行くのに、水上飛行機に乗ってます。わあ、すごい眺め！ 映画の中のワンシーンみたい！ これが写真。http://twitpic.com/xxx

sweet_pie01

@mari_tabi すごい！ 私もその水上飛行機乗ってみたい！ どんな感じか教えて。怖くない？

| **mari_tabi** Marippe | **kathleen_tl** Kathleen Taylor | **patty_coop** Patty Cooper | **sweet_pie01** Liu Fei |

Story 1 A Trip to Canada

mari_tabi

@sweet_pie01 It's not scary at all. I got a great view of the islands. It's a once in a lifetime experience, you should do it sometime!

> not 〜 at all：まったく〜でない　　once in a lifetime：一生に一度の
> experience：経験

mari_tabi

@kathleen_tl Hi, I just arrived at the harbour. I'm wearing a brown jacket and a blue hat. I think you can find me easily.

> easily：簡単に

kathleen_tl

@mari_tabi Welcome to Victoria! I'm at the entrance of the harbour right now waiting for Patty to arrive. She's coming by car.

> entrance：入口　　right now：現在、今　　wait for 〜：〜を待つ　　by car：車で

patty_coop

@mari_tabi @kathleen_tl Hey, I'm still in the parking lot. There are so many cars and it took longer than I thought, but I'll be there soon!

> parking lot：駐車場　　longer than I thought：思ってたより長く
> be there：そこに行く、着く

mari_tabi

@patty_coop Don't worry, take your time. I'm really happy you two are here to meet me!

> Don't worry.：心配しないで　　take one's time：時間をかける、ゆっくりする

mari_tabi

Wow, this is a great day! Kathleen and Patty are so nice. I'm really looking forward to going around town with them!

mari_tabi

@sweet_pie01 全然怖くないよ。島の景色がとっても素晴らしいの。一生に一度の経験ね、いつかやってみるといいわ！

mari_tabi

@kathleen_tl ハイ、港に着きました。茶色い上着を着て青い帽子をかぶってます。すぐに見つかると思うけど。

kathleen_tl

@mari_tabi ビクトリアへようこそ！ 今港の入り口でパティが着くのを待ってるの。彼女、車で来るのよ。

patty_coop

@mari_tabi @kathleen_tl ねえ、私まだ駐車場なの。車がいっぱいいて思ったより時間かかったけど、すぐに着くからね！

mari_tabi

@patty_coop 心配しないで、ゆっくりでかまわないよ。二人とも会いに来てくれてうれしい！

mari_tabi

ああ、いい1日だった！ キャスリーンもパティもすごくいい人。二人と一緒に街を回るのが、本当に楽しみ！

---

mari_tabi Marippe　**kathleen_tl** Kathleen Taylor　**patty_coop** Patty Cooper　**sweet_pie01** Liu Fei

Story 1 A Trip to Canada

DAY 18

mari_tabi

Today I'm going to visit Butchart Gardens with my friends in Victoria. One of them will drive us there. I'm so lucky!

one of them：彼らのうちの一人　　drive＋（人）：（人）を車で連れて行く

kathleen_tl

It's a nice day to visit a garden. I hope the roses are still blooming there. The Sunken Garden will be wonderful with all the roses.

bloom：咲く、花を開く　　Sunken Garden：サンクン・ガーデン（ブッチャート・ガーデンの一部。地面を掘った形になっている）

mari_tabi

It's funny, there's a "torii" in front of the Japanese garden. It's not exactly what we see in Japan, but it's still so beautiful!

funny：おかしい、面白い　　in front of 〜：〜の前に

kathleen_tl

The garden restaurant is so busy. Maybe we should try afternoon tea at the Fairmont Empress Hotel in town.

busy：混んでいる

mari_tabi

I could spend the whole day here at the Butchart Garden, but there's still so much to see in town. I wish I could stay longer!

18日目

mari_tabi

今日はビクトリアで友達とブッチャート・ガーデンに行くの。車で連れて行ってくれるんだって。ラッキー！

kathleen_tl

庭園に行くのにぴったりの日ね。まだバラが咲いているといいけれど。サンクン・ガーデンは、あのバラが咲いていると本当にきれいなのよね。

mari_tabi

面白いわね、日本庭園の真ん前に"鳥居"がある。日本で見るのとまったく同じじゃないけど、それでもすごくきれいだわ。

kathleen_tl

ガーデン・レストランはすごく混んでるわね。街のフェアモント・エンプレス・ホテルでアフタヌーン・ティーをした方がいいかも。

mari_tabi

ブッチャート・ガーデンに1日いてもいいけど、また街の中にも見る物がたくさんあるし。もっと長くいられたらいいのに！

| **mari_tabi** Marippe | **kathleen_tl** Kathleen Taylor | **patty_coop** Patty Cooper | **sweet_pie01** Liu Fei |

Story 1 A Trip to Canada

sweet_pie01

@mari_tabi I saw the pic of the garden. I love the one of the Sunken Garden. It's just like a scene in a dream!

scene：場面

mari_tabi

@sweet_pie01 You're right! You have to come here sometime. Now we are going back to the harbour to have afternoon tea!

You're right：そのとおり (rightは「正しい」)

sweet_pie01

@mari_tabi I'm going to start following your Canadian friends so I will be able to enjoy Canada more when I go there!

DAY 19

mari_tabi

I don't believe it! This is my last day in Canada! It was too short. I wish I had more time off. I will come here again!

kathleen_tl

Mari is leaving today. I'm going to pick her up with Patty at 8 am, so that she can make her flight.

so that ～：～できるように　　make ～：～に間に合う

mari_tabi

This time I'm taking a regular plane back to Vancouver and my friends are driving me to the airport. They've been so kind to me!

regular：普通の、普段の　　back to ～：～に戻る

A Trip to Canada **Story 1**

sweet_pie01

ブッチャート・ガーデンの写真見たよ。サンクン・ガーデンのやつがよかった。まるで夢の中の一場面みたいだね！

mari_tabi

@sweet_pie01 そうなの！ いつか来てみるといいわよ。これから港へ戻って、アフタヌーン・ティーなの！

sweet_pie01

@mari_tabi 私もあなたのカナダの友達をフォローすることにするわ。そうすれば、自分で行ったときに、もっとカナダを楽しめるでしょ！

19日目

mari_tabi

信じられない！ 今日がカナダで最後の日。短すぎ。もっと休みがあればいいのに。また来ようっと！

kathleen_tl

今日はマリが帰る日。飛行機に間に合うように、パティと8時に迎えに行くの。

mari_tabi

今回は、普通の飛行機でバンクーバーに戻るんだけど、友達が空港に連れて行ってくれるの。とっても親切にしてくれたわ！

---

mari_tabi	kathleen_tl	patty_coop	sweet_pie01
Marippe	Kathleen Taylor	Patty Cooper	Liu Fei

Story 1 A Trip to Canada

kathleen_tl

Everything seems OK. I'm leaving for the airport now. We really had a great time with Mari. I hope she can visit us again.

---

mari_tabi

I'm at the airport now. I have a lot of time here. I'll go to some Duty Free shops. Maybe I can find something interesting.

> Duty Free shop：免税店

mari_tabi

Oh, I haven't bought anything for my colleagues. Maybe these maple leaf cookies would be good, something typical of Canada.

> maple leaf：メープル・リーフ（カエデの葉。カエデは国旗にもデザインされているカナダの代表的な木で、カエデの葉の形をしたクッキーが有名なお土産）
> typical：典型的な、特有の

kathleen_tl

@mari_tabi I think you are already on the plane. I hope you enjoyed your stay in Canada. Let me know when you arrive in Tokyo.

---

kathleen_tl

I got a message from Tom! He's done with his project and he'll be on a flight to Vancouver tomorrow. I miss him so much!

> He's done with ～：彼は～を終えた（He's done＝He is done）
> miss ～：～に会いたい、会えなくて寂しい

A Trip to Canada **Story 1**

> **kathleen_tl**

全部OK。さてと、空港に行こう。マリと一緒で本当に楽しかったわ。また来てくれるといいんだけど。

> **mari_tabi**

空港にいます。ここでたくさん時間があるのよね。免税店にでも行こう。何か面白い物が見つかるかも。

> **mari_tabi**

あっ、同僚に何も買っていなかった。このメープル・リーフのクッキーがいいかも。カナダらしいでしょ。

> **kathleen_tl**

@mari_tabi もう飛行機に乗っているでしょう。カナダでの滞在を楽しんでくれたといいな。東京に着いたら教えてね。

> **kathleen_tl**

トムからメッセージが来た！ プロジェクトを終えたから、明日のバンクーバー行きの飛行機に乗るって。早く会いたい！

mari_tabi Marippe	**kathleen_tl** Kathleen Taylor	**patty_coop** Patty Cooper	**sweet_pie01** Liu Fei

Story 1 A Trip to Canada

DAY 20

mari_tabi

I'm back in Tokyo! I really had a great time in Canada. I hope I can go there again.

---

kathleen_tl

@mari_tabi I'm happy to hear that you arrived safely back in Tokyo. Please visit us again, and I'd love to visit Tokyo sometime!

> safely：無事に

patty_coop

@mari_tabi Back in Tokyo! It was great seeing you in Victoria. A lot of things to see, right? You could have spent a whole week here!

---

mari_tabi

I'm not sure if I can get back to the real world tomorrow. The trip to Canada was so impressive. I won't be able to forget it for a while.

> I'm not sure if ～：～かどうか分からない　　get back to ～：～に戻る
> impressive：印象が強い、感動的な

sweet_pie01

@mari_tabi Welcome back! Sounds like the trip to Canada was something special. Show me more pics, I'm looking forward to seeing them.

A Trip to Canada **Story 1**

20日目

mari_tabi

東京に戻ってきました！ カナダではとっても楽しかった。また行けるといいな。

kathleen_tl

@mari_tabi よかった、無事に東京に着いたのね。また来てね、私もいつか東京に行きたいわ！

patty_coop

東京に戻ったんだ！ ビクトリアで会えてよかった。見る物たくさんあったでしょ？ ここで1週間過ごしてもいいくらいよ！

mari_tabi

明日からまた現実の生活に戻れるかなあ。カナダ旅行、印象強かったから。しばらく忘れることができないだろうな。

sweet_pie01

@mari_tabiおかえりなさい！ カナダ旅行は何か特別なものだったみたいね。もっと写真見せて、楽しみにしているから。

| **mari_tabi** Marippe | **kathleen_tl** Kathleen Taylor | **patty_coop** Patty Cooper | **sweet_pie01** Liu Fei |

Story 1 A Trip to Canada

mari_tabi

@sweet_pie01 I took so many photos it will take time to sort them out. I'll unpack my luggage tomorrow. Good night!

sort ~ out：~を整理する　　unpack：荷を解く　　luggage：荷物

kathleen_tl

Mari is gone and Tom will be back soon. Do I need to go to the supermarket to get some food? Or are we going out for dinner?

be gone：行ってしまった

kathleen_tl

He'll be in Vancouver in 4 hours and he's taking me to a Chinese restaurant! I think I'll have spicy chicken noodles.

He'll be in Vancouver：バンクーバーに着くだろう（He'll＝He will）
spicy：辛い、香辛料の効いた　　noodle：ヌードル、麺

DAY 21

mari_tabi

I got so many things from Canada, but I should have bought more of these maple leaf cookies. They taste so good!

I should have bought：買うべきだった　　taste so good：とてもおいしい

mari_tabi

I uploaded some of my trip pics. The one I took from the seaplane is really amazing. Please take a look at it!
http://twitpic.com/xxx

upload：アップロードする

A Trip to Canada **Story 1**

mari_tabi
@sweet_pie01 写真たくさん撮ったから、整理するのに時間がかかりそう。荷物を開けるのは明日にしよう。おやすみ！

kathleen_tl
マリが行っちゃって、トムがすぐ戻ってくる。スーパーで食べ物を買ってきた方がいいかしら？ それとも、夕食は外で食べる？

kathleen_tl
トムが後4時間でバンクーバーに着く。チャイニーズ・レストランに連れて行ってくれるって！ ピリ辛チキンヌードルにしようかな。

21日目

mari_tabi
カナダですごくいろんな物を買ってきたけど、このメープル・リーフ・クッキー、もっと買ってくればよかったな。すごくおいしい！

mari_tabi
写真をいくつかアップロードしたの。水上飛行機から撮ったやつが、本当にすごい。ぜひ見てね！ http://twitpic.com/xxx

---

mari_tabi	kathleen_tl	patty_coop	sweet_pie01
Marippe	Kathleen Taylor	Patty Cooper	Liu Fei

Story 1 A Trip to Canada

kathleen_tl

@mari_tabi You just missed a chance to meet my boyfriend! He's back in Victoria now. I hope you are not suffering from jet lag.

> miss a chance：チャンスを逃す　　suffer from 〜：〜で苦しむ　　jet lag：時差ボケ

mari_tabi

@kathleen_tl Thank you so much for everything you did for me. I'm fine. Please visit me when you come to Japan!

patty_coop

@mari_tabi Since we've met on Twitter, I'm getting more interested in Japan. I hope I have a chance to visit your country sometime soon.

> since 〜：〜以来　　I'm getting more interested in 〜：〜にますます興味がわいてくる（「get＋比較級」で「ますます〜する」）

mari_tabi

@kathleen_tl @patty_coop It would be great if both of you could come over. I'll take you to see some real Japanese gardens in Kyoto!

> It would be great if 〜：〜だとうれしい

sweet_pie01

@mari_tabi I found Kathleen and Patty on Twitter. Now that I'm following them, we can talk about traveling together!

> now that 〜：今や〜なので　　traveling：旅行

mari_tabi

Wow, thanks to Twitter I know a lot of people all around the world now! Tweeting in English is so much fun!

> thanks to 〜：〜のおかげで　　all around the world：世界中
> so much fun：とても楽しいこと

A Trip to Canada **Story 1**

kathleen_tl

@mari_tabi私のボーイフレンドに会わせてあげることができなかったわね！ 今ビクトリアに帰ってます。あなたが時差ボケにかかってないといいけれど。

mari_tabi

@kathleenいろいろとありがとう。私は元気。日本に来るときは私の所に来てね！

patty_coop

@mari_tabi私たちがツイッターで会ってから、日本にもっと興味がわくようになってきたの。近いうちにあなたの国に行く機会があるといいな。

mari_tabi

@kathleen_tl @patty_coop二人とも来てくれたらうれしいな。京都の本物の日本庭園に連れて行ってあげる！

sweet_pie01

@mari_tabiツイッターでキャスリーンとパティ見つけたよ。私もフォローしてるから、一緒に旅行の話ができるね！

mari_tabi

わあ、ツイッターのおかげで、世界中のいろんな人たちと知り合いになれた！ 英語でつぶやくのって、楽しいね！

mari_tabi Marippe	**kathleen_tl** Kathleen Taylor	**patty_coop** Patty Cooper	**sweet_pie01** Liu Fei

> COLUMN 1

自分の意見や希望を伝える／依頼する表現

本書で紹介している英語表現は、普段の会話やメールでも活用できます。例えば、次のような言い方を覚えておくといいでしょう。

I'd like to know about ～ (p.8)　～について知りたいのですが

I'd love to ～ (p42)　ぜひ～したいんです

I hope ～ (p10)　～だといいんですが

It'd be great if ～ (p28)　～したら素晴らしいですね

これらは、会話の中で **I want to ～**（～したい）を丁寧に言いたいというときに使えます。また、**It'd be great if you can help us.**（助けていただけるとありがたいのですが）といったように、婉曲な依頼の表現としても使えます。ここで、依頼の表現もまとめておきましょう。

Can you ～?　～してもらえますか？

Could you ～?　～していただけますか？

Is it possible for you to ～?　～することは可能でしょうか？

I'd appreciate it if you could ～　～していただけるとありがたいのですが

I'd appreciate it if you could ～は一見丁寧すぎるようにも思えるかもしれませんが、ビジネスの場面では意外によく見かけます。英語でメールを書くときなど、ぜひ一度使ってみてください。

Story 2
My Company's Official Language will be English!

おやじのグローバル化
(T_T) 泣き言つぶやき

Kenji Matsumoto

Kenji Matsumoto
@KenMatsuGG Tokyo, Japan
男性。43歳。IT企業のエンジニア。妻と小学生の子どもがいる

I'm Japanese. I work for a computer company as an engineer. I don't speak English very well but I'm trying to improve.

僕は日本人です。コンピューター会社でエンジニアとして働いています。英語はあまり上手に話せませんが、上達しようとしています。

Micheal J Chen
@MJ_CoolChen Singapore
男性。中国系のシンガポール人。金融関係の会社に勤める。最新のIT機器が好き。Kenjiと気が合う

Financial analyst, single, ambitious, devoted to my career and video games!

金融アナリスト、独身、野心家、仕事と、そしてテレビゲームに情熱を捧げてます！

Sarah Reynolds
@Sarah2000ry Tokyo, Japan
女性。アメリカ人。日本在住英語教師。Kenjiに英会話のアドバイスをくれる

Language teacher living in Japan. Enjoying life here and looking for new friends.

日本に住んでいる語学教師です。ここでの生活を楽しんでいて、新しい友人を探しています。

Story 2 My Company's Official Language will be English!

DAY 1

KenMatsuGG

Hello. I'm an engineer at a computer company in Tokyo, I'm married and we have a 9 year old son. I hope you can understand my English.

MJ_CoolChen

@KenMatsuGG Hi. I've started following you. No problem with the English. What kind of computers do you work on?

KenMatsuGG

@MJ_CoolChen Thank you for following me. I check system errors. The job is not very easy, but I enjoy it.

easy：やさしい　　enjoy：楽しむ

MJ_CoolChen

I like computer games. And actually, I was thinking about working in Tokyo. Any thoughts on the financial jobs there?

actually：実は　　thought：考え　　financial：金融の

KenMatsuGG

@MJ_CoolChen There are many financial companies here. Where do you live? What does your family think about your job?

MJ_CoolChen

@KenMatsuGG I live in Singapore, just by myself. I think Tokyo would be an exciting place to work. Or maybe Shanghai.

by myself：一人で　　exciting：面白い、わくわくする
or：もしくは、さもなければ　　Shanghai：上海

1日目

KenMatsuGG

こんにちは。僕は東京のコンピューター会社のエンジニアです。結婚していて、9歳の息子がいます。僕の英語を分かってもらえるといいのですが。

MJ_CoolChen

@KenMatsuGG ハイ。フォロー始めました。英語大丈夫ですよ。コンピューターのどんな仕事をしているんですか？

KenMatsuGG

@MJ_CoolChen フォローしてくれてありがとう。僕はシステム・エラーのチェックをしています。仕事はそんなにやさしくありませんが、僕は楽しんでやっています。

MJ_CoolChen

僕はコンピューター・ゲームが好きなんだ。それから実は、東京で働こうかと考えてる。そっちでの金融関係の仕事について、どう思う？

KenMatsuGG

@MJ_CoolChen 金融関係の会社はたくさんあります。あなたはどこに住んでいるんですか？ 家族の人は、あなたの仕事についてどう思っているんですか？

MJ_CoolChen

@KenMatsuGG 僕はシンガポールに一人で住んでいるんだ。東京は働きがいのありそうな街だよね。それか、上海とか。

KenMatsuGG
Kenji Matsumoto

MJ_CoolChen
Micheal J Chen

Sarah2000ry
Sarah Reynolds

Story 2 My Company's Official Language will be English!

KenMatsuGG

Tokyo is a safe, clean city. You'll like it. I haven't been to Shanghai but I think it's a nice city too. RT **@MJ_CoolChen** Or maybe Shanghai.

MJ_CoolChen

I got a message from my girlfriend. She'd like to have dinner with me tonight. She wants something spicy. What should I do?

> something spicy：何か辛い物（spicyは「辛い、香辛料の効いた」）

DAY 2

KenMatsuGG

I heard our company made a big decision. English will be our company's official language. Everyone in my department is nervous.

> big decision：重大な決断（decisionは「決断」）　　official language：公用語
> department：部署　　nervous：慌てる、イライラする

MJ_CoolChen

@KenMatsuGG Don't be nervous. Look at it as a challenge. You will be able to work anywhere in the world with good English skills.

> look at ～ as …：～を…だと思う、考える　　challenge：チャレンジ、挑戦
> anywhere：どこででも　　skill：スキル、技術

KenMatsuGG

@MJ_CoolChen Thank you. But I don't think my English is good enough. Can you tell me how to improve my English?

> good enough：十分良い、上手な　　how to improve：上達させ方、上手になる方法

> **KenMatsuGG**

東京は安全で清潔な街です。きっと気に入ります。上海には行ったことがありませんが、そこもいい所だと思います。RT **@MJ_CoolChen** それか、上海とか。

> **MJ_CoolChen**

彼女からメールだ。今晩一緒に食事したいって。何か辛い物がいいのか。どうしようか？

2日目

> **KenMatsuGG**

僕たちの会社は重大な決断をした。英語が会社の公用語になるのだ。僕たちの部署の人は皆、焦っている。

> **MJ_CoolChen**

@KenMatsuGG 焦ることない。チャレンジだと思えばいいよ。英語のスキルがあれば、世界中どこに行っても働けるからね。

> **KenMatsuGG**

@MJ_CoolChen ありがとう。でも、僕は自分の英語が十分だとは思っていません。どうしたら英語が上達するか、教えてもらえますか？

| **KenMatsuGG** Kenji Matsumoto | **MJ_CoolChen** Micheal J Chen | **Sarah2000ry** Sarah Reynolds |

Story 2 My Company's Official Language will be English!

MJ_CoolChen

@KenMatsuGG Just speak and use English whenever you can. You can even practice by trying to think in English.

> whenever 〜：〜のときはいつでも　　even 〜：〜さえ
> practice：練習する　　think in English：英語で考える

KenMatsuGG

I had lunch with my coworkers today. We talked about how to study English, and I heard Twitter is useful to get information.

> coworker：同僚　　useful：役に立つ

KenMatsuGG

I'll look for English conversation schools and other ways to study English on the Internet. I hope I can find a good way.

> look for：探す
> English conversation school：英会話学校（conversationは「会話」）

MJ_CoolChen

Wow! A friend says the economy in Dubai is taking a turn for the worse again. I'm shocked because everyone here thought it would be OK.

> economy：経済　　Dubai：ドバイ（アラブ首長国連邦の都市。中東の金融・経済の一大中心地）　　take a turn for the worse：悪化する（worseは「より悪い」）
> shocked：ショックを受けた

MJ_CoolChen

Here in Singapore the economy is not contracting as much as people thought. Many analysts say it's doing well.

> contract：（経済が）縮小する　　analyst：分析家、アナリスト

MJ_CoolChen

@KenMatsuGG できるときはいつでも英語を話して英語を使えばいいんだよ。英語で考えてみることで練習することだってできるんだ。

KenMatsuGG

今日同僚と昼食を取った。英語の勉強の仕方について話をして、ツイッターは情報を得るのに役立つと聞いた。

KenMatsuGG

インターネットで、英会話学校やほかの英語の学習方法を探そう。いい勉強方法が見つかるといいのだけれど。

MJ_CoolChen

おっと！ 友人が、ドバイの経済はまた悪化すると言っている。ショックだな、こっちでは皆大丈夫だろうと思ってたから。

MJ_CoolChen

ここシンガポールでは、人が思うほどには経済は縮小していない。うまくいっているという専門家はたくさんいる。

KenMatsuGG
Kenji Matsumoto

MJ_CoolChen
Micheal J Chen

Sarah2000ry
Sarah Reynolds

Story 2 My Company's Official Language will be English!

DAY 3

KenMatsuGG

I didn't know English schools cost so much. I don't think I can afford it. I have to pay my mortgage and other bills.

> cost：お金がかかる　　afford ～：～を買う余裕がある　　pay：支払う
> mortgage：住宅ローン　　bill：請求書、勘定書

MJ_CoolChen

@KenMatsuGG Even though the price might be a bit high, it is like an investment in your future, don't you think?

> even though ～：たとえ～でも　　a bit：少し、ちょっと
> investment：資本　　future：未来

KenMatsuGG

@MJ_CoolChen I understand that it is good for me, but my son is going to Juku. I don't know how my wife will react to another bill.

> react to ～：～に反応する

KenMatsuGG

There must be a cheaper way to study English. I could use some books and start a study group with some people at work.

> there must be ～：～があるに違いない　　could ～：～できるだろう
> study group：学習グループ　　at work：仕事（場）で

MJ_CoolChen

@KenMatsuGG In Singapore we begin learning English when we are little children. I thought it was the same in Japan. What's Juku?

> the same in Japan：日本でも同じ

3日目

KenMatsuGG

英語学校がこんなにお金かかるとは知らなかった。そんな余裕があるとは思えないな。住宅ローンやほかのお金を払わないといけないんだ。

MJ_CoolChen

@KenMatsuGG 多少料金が高くても、将来への投資のようなものでしょ、そう思わない？

KenMatsuGG

@MJ_CoolChen 自分にとっていいということは分かりますが、息子は塾に行ってます。また新たにお金がかかることに、うちの奥さんがどう反応するか分かりません。

KenMatsuGG

もっと安く英語を勉強する方法があるに違いない。本を使って、職場の人たちと学習グループを始めるのがいいかもしれない。

MJ_CoolChen

@KenMatsuGG シンガポールでは、小さい子どものころから英語を学び始めるんだ。日本でも同じだと思ってた。Jukuって何？

KenMatsuGG — Kenji Matsumoto

MJ_CoolChen — Micheal J Chen

Sarah2000ry — Sarah Reynolds

Story 2 My Company's Official Language will be English!

KenMatsuGG

@MJ_CoolChen Children study at Juku after school. Japanese children study English in elementary schools now, but not when I was in school.

> regular：普通の、いつもの　　elementary school：小学校

MJ_CoolChen

Rachel wants to go out for dinner again tonight? We are seeing each other more often lately. I think it's a little too much.

> go out for dinner：夕食に出かける　　see：会う　　more often：より頻繁に
> lately：最近　　too much：多過ぎる

MJ_CoolChen

I told her I have to work overtime tonight. Actually I have a report to finish and a late meeting with my manager.

> work overtime：残業する　　finish：終える　　late：遅い
> meeting：打ち合わせ、会議　　manager：責任者、マネジャー

DAY 4

KenMatsuGG

I found many people studying English on Twitter. I'll start following them to find more ways to study.

> follow：フォローする

MJ_CoolChen

@KenMatsuGG You should follow other engineers on Twitter. It's a good way to start studying English since you are in that field.

> since：〜なので　　field：分野、領域

KenMatsuGG
@MJ_CoolChen 子どもたちは学校の後、塾で勉強します。日本人の子どもたちは今は小学校で英語を勉強していますが、私が学校にいたころは違いました。

MJ_CoolChen
レイチェルは今晩も食事に出たいって？ 最近、会う頻度が高くなってるな。ちょっと多過ぎるかもしれない。

MJ_CoolChen
今晩は残業しないといけないって言った。実際、終えなければならない報告書があるし、責任者との打ち合わせもあるんだ。

4日目

KenMatsuGG
ツイッターで英語を勉強している人をたくさん見つけた。もっといろいろな勉強方法を見つけるために、この人たちをフォローしてみよう。

MJ_CoolChen
@KenMatsuGG ツイッターでほかのエンジニアをフォローするといいよ。英語を勉強し始めるのにいいよ、君はその分野の仕事をしているんだから。

KenMatsuGG Kenji Matsumoto

MJ_CoolChen Micheal J Chen

Sarah2000ry Sarah Reynolds

Story 2 My Company's Official Language will be English!

KenMatsuGG

@Sarah2000ry Hello. Are you teaching English in Tokyo? I need to practice English. My company will make it the official language in 2 years.

practice：練習する　make it ～：それを～にする　in 2 years：2年後に

Sarah2000ry

@KenMatsuGG Thanks for signing up to follow me. I'm sure I can help you. Ask me anything you want anytime.

sign up：登録する　I'm sure ～：きっと～だと思う
anything：何でも　anytime：いつでも

MJ_CoolChen

Just got the newest Computer Game Magazine and can't wait to see what new games are coming out.

newest：最新の　come out：出てくる、発売される

MJ_CoolChen

Listen to this, it's fantastic! Mantendo is launching a new game console. I've been waiting for this news.

fantastic：すごい、素晴らしい　launch：売り出す
game console：ゲーム機（consoleは「制御盤、コンソール」）

MJ_CoolChen

@KenMatsuGG Do you know anything about the new Mantendo console? It won't be out here for a couple of months.

won't be out：発売されない　a couple of ：2、3の

My Company's Official Language will be English! **Story 2**

KenMatsuGG

@Sarah2000ry こんにちは。東京で英語を教えているんですか？ 僕は英語を練習する必要があるんです。うちの会社は、2年後に英語を公用語にするんです。

Sarah2000ry

@KenMatsuGG 私をフォローしてくれてありがとう。きっとお手伝いできると思います。いつでも何でも聞いてちょうだい。

MJ_CoolChen

最新の『コンピューター・ゲーム・マガジン』を手に入れた。どんな新しいゲームが出てくるか、待ちきれないな。

MJ_CoolChen

聞いてくれ、すごいぞ！ マンテンドーが新しいゲーム機を出すんだ。この知らせを待ってたんだよ。

MJ_CoolChen

@KenMatsuGG マンテンドーの新しいゲーム機について、何か知ってる？ ここでは後2、3カ月しないと発売しないんだけれど。

KenMatsuGG
Kenji Matsumoto

MJ_CoolChen
Micheal J Chen

Sarah2000ry
Sarah Reynolds

Story 2 My Company's Official Language will be English!

KenMatsuGG

@MJ_CoolChen Do you mean the new game machine? I could get some information, but in Japanese. I think you can find English news online.

> online：オンラインで、インターネットで

DAY 5

KenMatsuGG

We tried an all English meeting this morning. Some people could speak well, but others couldn't. It was very strange.

> all English meeting：すべて英語で行う会議
> some 〜, others …：ある人は〜で、そのほかの人は…

MJ_CoolChen

@KenMatsuGG My company is very mixed. Chinese, Indian and some Brits. Our meetings are out of control, but it works out in the end.

> mixed：混ざった　　Brit：イギリス人（Britishの口語的な言い方）
> out of control：コントロールがきかない、収拾つかない　　work out：うまくいく
> in the end：結局、最後には

KenMatsuGG

@MJ_CoolChen Our meetings usually take a long time. And if we have to use English, we'll be in meetings forever.

> usually：普段、通常　　take a long time：とても時間がかかる

KenMatsuGG

@MJ_CoolChen 新しいゲームマシンのこと？ 情報は得られるけれど、日本語です。オンラインで英語のニュースがあると思いますよ。

5日目

KenMatsuGG

けさは全部英語の会議をやってみた。上手に話せる人もいたけれど、できない人もいた。ヘンな感じだった。

MJ_CoolChen

@KenMatsuGG うちの会社はいろいろ混じってるんだ。中国人、インド人に、イギリス人。うちの会議は収拾つかないけれど、最後にはうまくいくよ。

KenMatsuGG

@MJ_CoolChen うちの会議は、普段すごく時間がかかります。英語を使わなければならないとすると、永久に続いてしまうかも。

KenMatsuGG — Kenji Matsumoto

MJ_CoolChen — Micheal J Chen

Sarah2000ry — Sarah Reynolds

Story 2　My Company's Official Language will be English!

KenMatsuGG

It took us almost 1 hour to decide what to discuss at today's meeting. Then we only had a half an hour for the topic.

> take +（人）+（時間）to 〜：（人）にとって〜するのに（時間）かかる
> decide：決める、決心する　　what to discuss：何を話し合うか（discussは「話し合う」）
> a half an hour：30分　　topic：トピック、テーマ

MJ_CoolChen

@KenMatsuGG Sounds like you need to get more organized. Don't worry, it will get better.

> more organized：より整理された、より組織立った　　worry：心配する
> get better：よくなる

MJ_CoolChen

My friend doesn't think the Dubai economy will pick up very soon, especially since the US dollar should be weak for a while.

> economy：経済　　pick up：（景気などが）回復する、よくなる　　soon：すぐに
> especially：特に　　since 〜：〜だから　　for a while：しばらく

MJ_CoolChen

I told him we can still make money in a bad economy. It's not as easy as in the good times. We have to be creative.

> still：依然として、それでも　　make money：お金をもうける
> bad economy：不景気　　as easy as 〜：〜ほど簡単ではない
> creative：工夫している、独創性がある

Sarah2000ry

@KenMatsuGG Hello! How's your English studying? I always suggest that people check my old posts if they want to learn English.

> suggest：示唆する、提案する　　post：（ツイッターへの）投稿

KenMatsuGG

今日の会議で何を話し合うか決めるのに、ほとんど1時間かかった。それからテーマについて話すのに30分しかなかった。

MJ_CoolChen

@KenMatsuGG もっとやり方を整理した方がいいようだね。心配しないで、きっとよくなるよ。

MJ_CoolChen

友人は、ドバイの経済がそんなにすぐに回復するとは思っていない、特にUSドルがしばらく弱いだろうから。

MJ_CoolChen

彼に、不景気のときでも金もうけはできるって言ってあげたんだ。いい時期ほど簡単じゃない。工夫しないとね。

Sarah2000ry

@KenMatsuGG こんにちは！ 英語の勉強はどう？ 英語の勉強をしたいなら、私の前の投稿を見てみるよう、いつも言ってるの。

KenMatsuGG — Kenji Matsumoto

MJ_CoolChen — Micheal J Chen

Sarah2000ry — Sarah Reynolds

Story 2 My Company's Official Language will be English!

DAY 6

KenMatsuGG

I wanted to take a midday nap today, but my son won't let me relax. I guess I'll take him to the park.

```
midday nap：昼寝    let+（人）+（動詞）：（人）に〜させる
relax：リラックスする、ゆっくりする    guess 〜：〜と思う
take+（人）+to 〜：（人）を〜へ連れて行く
```

KenMatsuGG

My son is playing baseball with some other kids in the park. I can do some studying while I wait for him.

```
kid：子ども（childの口語的な言い方）    while 〜：〜の間    wait for 〜：〜を待つ
```

MJ_CoolChen

I'm meeting Rachel in front of Dean's Shoe Store. She wants to check a pair of shoes before we go to a movie.

```
in front of 〜：〜の前
```

MJ_CoolChen

Shopping is so boring, but Rachel loves it. She's trying on shoes, so I'm checking tweets from my iPhone.

```
boring：退屈な    try on：試す、試着する
```

KenMatsuGG

My wife wants to go on a trip during the spring holiday. I can't think about that because I'm worried about studying English.

```
go on a trip：旅行に行く    spring holiday：春休み
be worried about 〜：〜を心配する
```

6日目

KenMatsuGG

今日は昼寝したかったけれど、息子がゆっくりさせてくれない。公園にでも連れて行こうか。

KenMatsuGG

息子は公園でほかの子どもと野球をしている。待っている間、いくらか勉強できるな。

MJ_CoolChen

ディーン靴店の前で、レイチェルと待ち合わせ。映画に行く前に、靴を見てみたいそうだ。

MJ_CoolChen

買い物はすごく退屈だけど、レイチェルは買い物が大好き。彼女が靴を試しているから、iPhoneでツイートを見てるんだ。

KenMatsuGG

うちの奥さんは、春休みに旅行に行きたいそうだ。僕は英語の勉強が心配で、そのことについては考えられないな。

KenMatsuGG
Kenji Matsumoto

MJ_CoolChen
Micheal J Chen

Sarah2000ry
Sarah Reynolds

Story 2 My Company's Official Language will be English!

MJ_CoolChen

I'd like to go to the computer store, but Rachel says I waste too much time with those silly games. She doesn't get it.

> waste：ムダにする　　too much：あまりにも多くの　　silly：くだらない、ばかげた
> get it：分かる、理解する

KenMatsuGG

I have to talk about studying English with my wife. I really need to do it. I hope she understands.

MJ_CoolChen

Rachel asked me about some shoes she picked out. I have no idea what the difference is. Looks like I don't get it either - lol

> pick out：選ぶ　　have no idea：分からない　　look like 〜：〜のようだ、〜のように見える　　〜 not ... either：〜も…でない　　lol：大笑い（＝laughing out loud。日本語ツイッターではよくwが使われる）

DAY 7

Sarah2000ry

@KenMatsuGG Have you considered private lessons? I offer lessons and I'm pretty flexible about schedule and locations.

> consider：考える　　private lesson：プライベート・レッスン
> offer：申し出る、提案する　　pretty：とても　　flexible：フレキシブルな、柔軟な
> location：場所

KenMatsuGG

@Sarah2000ry I checked some schools near me and the fees are quite high. Can you tell me how much you charge?

> fee：料金　　quite：とても　　charge：請求する

MJ_CoolChen

コンピューター・ショップに行きたいんだけれど、レイチェルは僕が、そういうくだらないゲームであまりにも多くの時間をムダにしているって言う。分かってないな。

KenMatsuGG

うちの奥さんと、英語の勉強について話をしないと。本当にやらないといけないんだ。分かってくれるといいけれど。

MJ_CoolChen

レイチェルが僕に、選んだ靴について聞いてきた。僕には違いが分からない。僕も分かってないみたいだなw

7日目

Sarah2000ry

@KenMatsuGG プライベート・レッスンについて考えてみたことある？ 私はレッスンをしているの。スケジュールも場所も、とてもフレキシブルにできるわよ。

KenMatsuGG

@Sarah2000ry 近くの学校をチェックしてみたら、授業料がとても高かったんです。あなたの料金はいくらか教えてくれますか？

KenMatsuGG Kenji Matsumoto　　**MJ_CoolChen** Micheal J Chen　　**Sarah2000ry** Sarah Reynolds

Story 2 My Company's Official Language will be English!

MJ_CoolChen

Couldn't wait for Mantendo's new console, so I got the new Zonie-P10. Just opened the box!

MJ_CoolChen

Wow. It's a little heavier than I thought but it feels great in my hand. I wonder how long the battery will last.

> heavier：より重い（heavyの比較級）　　I wonder 〜：〜だろうかと思う
> how long：どれくらいの時間　　battery：バッテリー、電池　　last：もつ、続く

Sarah2000ry

@KenMatsuGG Depending on the class style, I charge from 2500-5000 per hour. It will also depend on the length and where we do lessons.

> depend on 〜：〜による　　per hour：1時間ごとに　　length：長さ
> where we do 〜：〜をする場所

KenMatsuGG

@Sarah2000ry I'll think about my schedule. I will let you know when I am available later. Then we can discuss more details. Thanks.

> available：都合がつく、スケジュールが空いている　　later：後で
> detail：細部、詳細

KenMatsuGG

@Sarah2000ry I realized writing tweets is a good way to learn English. Please tell me if my English is strange.

> realize 〜：〜と分かる、気付く　　good way to 〜：〜するのにいい方法

MJ_CoolChen

マンテンドーの新しいゲーム機が出るのが待てなくて、新しいZonie-P10を買った。箱を開けたところだ！

MJ_CoolChen

おっと。思ってたより重いけど、手に持った感じはいいな。バッテリーはどれくらいもつんだろう。

Sarah2000ry

@KenMatsuGG クラスのスタイルによるけれど、1時間2500～5000円いただいています。レッスンの長さやレッスンをする場所にもよるわ。

KenMatsuGG

@Sarah2000ry スケジュールを考えてみます。後で、いつが都合がいいか知らせます。それからもっと細部について話し合いましょう。ありがとう。

KenMatsuGG

@Sarah2000ry つぶやきを書くのは、英語を勉強するのにいいんだと分かりました。僕の英語がヘンだったら教えてください。

KenMatsuGG
Kenji Matsumoto

MJ_CoolChen
Micheal J Chen

Sarah2000ry
Sarah Reynolds

Story 2　My Company's Official Language will be English!

MJ_CoolChen

3 hours of gaming. The battery is still going strong, but I'm wiped out. Love this new machine though, highly recommended.

go strong：調子がいい、元気でいる　　wiped out：疲れ切った
though：だけど、でも　　highly：非常に、大変　　recommended：お薦めの

DAY 8

KenMatsuGG

We had another briefing about working in English at the office today. A lot of people in the company must be worried.

another：新たな、もう一つの　　briefing：状況説明、報告

KenMatsuGG

I'm doing a search for English learning podcasts. They said, "Listening to English for 1-hour a day is a good idea."

do a search：調べ物をする、検索する　　podcast：ポッドキャスト　　a day：1日に

KenMatsuGG

I found so many podcasts. How can I decide which are the best? I'll just have to try some of them.

MJ_CoolChen

@KenMatsuGG Check CNN or other news site podcasts. You can keep up with current events and study at the same time.

keep up with ～：～についていく　　current event：最近の出来事、時事問題
at the same time：同時に

MJ_CoolChen
3時間ゲーム。バッテリーはまだもってるけど、疲れ切った。でもこの新しいゲーム機は気に入った、すごくオススメ。

8日目

KenMatsuGG
今日は、職場で英語を使って働くことについて、また説明があった。会社の多くの人は、心配しているに違いない。

KenMatsuGG
英語を勉強するためのポッドキャストについて調べてみた。「1日に1時間英語を聞くとよい」と書いてある。

KenMatsuGG
ポッドキャストがすごくたくさん見つかった。どれが一番いいか、どうやって決めればいいんだろう？ いくつか聞いてみないと。

MJ_CoolChen
@KenMatsuGG CNNやほかのニュースサイトのポッドキャストをチェックするといいよ。最近の出来事についていくことができると同時に、勉強もできるんだよ。

KenMatsuGG
Kenji Matsumoto

MJ_CoolChen
Micheal J Chen

Sarah2000ry
Sarah Reynolds

Story 2 My Company's Official Language will be English!

MJ_CoolChen

I just heard my boss is leaving. She found a new job in Shanghai. I am a little jealous.

leave：去る、いなくなる　　jealous：うらやましがる、嫉妬する

MJ_CoolChen

I guess we'll have a new boss soon. I wonder if they will promote from within, or bring in someone from another company.

promote：昇進させる　　within：内部　　bring in 〜：〜を連れてくる、呼び寄せる

KenMatsuGG

@MJ_CoolChen Your new boss will probably be nice. By the way, I will check some CNN podcasts like you said. I hope I can understand them.

probably：多分、恐らく　　by the way：ところで
like you said：あなたが言ったように

Sarah2000ry

@KenMatsuGG CDs and podcasts can be good, but sometimes they're a little too hard for beginners. I think English radio programs are great.

sometimes：ときどき　　beginner：初級者、初心者　　radio program：ラジオ番組

DAY 9

KenMatsuGG

@Sarah2000ry I've been using CNN podcasts but they are too fast for me. I'll try an NHK radio English program instead.

MJ_CoolChen

上司が辞めるって聞いた。上海で新しい仕事を見つけたそうだ。ちょっとうらやましいな。

MJ_CoolChen

すぐに新しい上司が来るだろう。社内から昇進させるか、それともほかの会社から誰か連れてくるのかな。

KenMatsuGG

@MJ_CoolChen 新しい上司は、多分いい人ですよ。ところで、あなたが言ったように、CNNのポッドキャストを聞いてみることにします。僕に分かるといいんですが。

Sarah2000ry

@KenMatsuGG CDやポッドキャストはいいけれど、初級者にはちょっと難しすぎることもあるわ。私は、英語のラジオ番組がいいと思うの。

9日目

KenMatsuGG

@Sarah2000ry CNNのポッドキャストを使ってきましたが、僕には速すぎるんです。代わりにNHKラジオの英語番組を聴いてみることにします。

KenMatsuGG Kenji Matsumoto

MJ_CoolChen Micheal J Chen

Sarah2000ry Sarah Reynolds

Story 2 My Company's Official Language will be English!

Sarah2000ry

@KenMatsuGG NHK is good, but you don't have to give up on the CNN podcasts. Your ear will get accustomed to the native speakers' speed.

> give up on 〜：〜を止める、あきらめる　　get accustomed to 〜：〜に慣れる
> native speaker：ネイティブ・スピーカー

MJ_CoolChen

The new boss is here, an Indian man that used to live in Tokyo, near Shibuya. He seems very serious.

> used to 〜：以前〜していた　　seem 〜：〜のようだ　　serious：まじめな

MJ_CoolChen

I was right. I made a joke at a meeting and the new boss gave me a strange look. Work will be tough if we can't get along.

> right：正しい、合っている　　make a joke：ジョークを言う
> give 〜 a strange look：〜をヘンな目で見る、けげんな顔をする　　tough：大変な
> get along：仲良くする、うまくやっていく

KenMatsuGG

@Sarah2000ry Thanks for the advice. Do you have time to meet me the day after tomorrow to discuss lessons? Around 6:30 near Tokyo Station?

Sarah2000ry

@KenMatsuGG Sure, I'm free. I'll send my cell# in case you need to call. I'll bring some materials for a trial and level check lesson.

> free：用事がない、(時間が) 空いている　　cell#：携帯番号 (cellはcell phone [携帯電話] のこと、#はnumberの略号)　　in case 〜：〜に備えて　　material：教材
> trial：トライアル、試験

Sarah2000ry

@KenMatsuGG NHKはいいけれど、CNNのポッドキャストをあきらめる必要はないわ。耳がネイティブ・スピーカーのスピードに慣れるわよ。

MJ_CoolChen

新しい上司が来た、前に東京の渋谷近くに住んでいたことがあるインド人だ。すごくまじめそうだな。

MJ_CoolChen

やっぱり。会議でジョークを言ったら、新しい上司にヘンな目で見られた。上司とうまくいかなかったら、仕事やりづらいだろうな。

KenMatsuGG

@Sarah2000ry アドバイスをありがとう。レッスンについて話し合うために、あさって会う時間はありますか？ 6時半ごろ、東京駅の近くは？

Sarah2000ry

@KenMatsuGG いいわよ、空いてます。電話する必要が出たときのために、携帯の番号を教えておくわね。トライアルやレベルチェックレッスンのための教材を持っていくわ。

KenMatsuGG — Kenji Matsumoto

MJ_CoolChen — Micheal J Chen

Sarah2000ry — Sarah Reynolds

Story 2 My Company's Official Language will be English!

MJ_CoolChen

Rachel is at the airport. She's going to Russia for business. She'll be gone for a week.

for business：仕事で　　be gone：いなくなる

MJ_CoolChen

I really don't like my new boss very much. I hope I still have a job when Rachel comes back. lol

DAY 10

Sarah2000ry

@KenMatsuGG See you tomorrow outside the Marunouchi North gate. I'll be wearing a white hat and carrying a purple bag.

outside ～：～の外側　　～ gate：～口　　carry：持っている、運ぶ

KenMatsuGG

@MJ_CoolChen I will meet an English teacher tomorrow, but I'm a bit nervous about talking to her.

a bit：ちょっと、少し　　nervous：不安な、緊張した

KenMatsuGG

@MJ_CoolChen I hope I can explain what I want to study. I suppose she should understand, since she is an English teacher.

explain：説明する　　suppose ～：～だと思う、推測する　　should ～：～のはずだ

MJ_CoolChen

レイチェルは空港にいる。仕事でロシアに行くそうだ。1週間いなくなる。

MJ_CoolChen

本当に新しい上司が気に入らない。レイチェルが帰ってきたときに、まだ僕の仕事があるといいけれどw

10日目

Sarah2000ry

@KenMatsuGG 明日は丸の内北口の外で会いましょう。私は白い帽子をかぶって、紫のバッグを持っています。

KenMatsuGG

@MJ_CoolChen 明日英語の先生に会うんだけれど、話をするのが、ちょっと不安なんです。

KenMatsuGG

@MJ_CoolChen 何を勉強したいか、説明できるといいんだけれど。英語の先生だから、分かってくれるはずですよね。

KenMatsuGG Kenji Matsumoto	**MJ_CoolChen** Micheal J Chen	**Sarah2000ry** Sarah Reynolds

Story 2 My Company's Official Language will be English!

MJ_CoolChen

The new boss is making us do weekly progress reports. What a pain! I've never had to do that before.

> make+(人)+(動詞):(人)に〜させる　weekly progress report:週間進捗報告書
> What a pain!:面倒くさい!(painは「苦痛」)

MJ_CoolChen

He's worked in a few countries, but he really doesn't understand our culture yet. I'm really not very happy.

> a few:いくつかの、2、3の　culture:カルチャー、文化　not yet:まだ

KenMatsuGG

@MJ_CoolChen I suppose it's difficult to work for an international company because there are so many differences among employees.

> among 〜:〜の間で　employee:社員、従業員

Sarah2000ry

@KenMatsuGG Sorry, but I have to change the time we meet. I can't get to the station until 6:45. I hope that's OK.

KenMatsuGG

@Sarah2000ry No problem. I can go to a coffee shop or bookstore to wait. I will see you at 6:45. Thanks.

> coffee shop:コーヒーショップ、喫茶店　bookstore:本屋

MJ_CoolChen

新しい上司は僕らに週間進捗報告書を書かせる。面倒くさい！ 前はやらなくてもよかったのに。

MJ_CoolChen

彼はいくつかの国で働いてきたけれど、まだ僕らのカルチャーをまるで分かっていない。本当に気に入らないな。

KenMatsuGG

@MJ_CoolChen 国際企業では社員の間に違いがたくさんあるから、仕事をするのが難しいのでしょうね。

Sarah2000ry

@KenMatsuGG ごめんなさい、待ち合わせ時間を変えないといけないの。駅に着くのは6時45分になります。大丈夫だといいのですが。

KenMatsuGG

@Sarah2000ry 大丈夫です。コーヒーショップか本屋に行って待っています。6時45分に会いましょう。ありがとう。

KenMatsuGG
Kenji Matsumoto

MJ_CoolChen
Micheal J Chen

Sarah2000ry
Sarah Reynolds

Story 2　My Company's Official Language will be English!

DAY 11

MJ_CoolChen

Busy at work now. Things are all different with the new boss and I'm still not used to everything.

> things are all different：何もかもが（前とは）異なる　　be used to ～：～に慣れる

KenMatsuGG

I found out how to upload pictures. Check out the photo of my last family trip. We went to a hot spring in Izu.
http://twitpic.com/xxx

> find out：発見する、分かる　　upload：アップロードする　　check out ～：～を見てみる、確認する　　family trip：家族旅行　　hot spring：温泉

KenMatsuGG

Here's another picture of my son playing baseball. He really loves it and wants to be a professional player someday.
http://twitpic.com/xxx

> professional：プロの　　someday：いつか

MJ_CoolChen

@KenMatsuGG Nice family pics. You look very happy in the vacation shot.

> pic=picture：写真　　vacation：休暇、休日　　shot：（写真の）場面

KenMatsuGG

I finally spoke to my wife about the company's plan for English as the official language. She wasn't very interested.

> plan：計画、予定　　interested：興味がある

MJ_CoolChen

11日目

MJ_CoolChen

今、仕事で忙しい。新しい上司の下で何もかもが違い、僕はまだ全然慣れてない。

KenMatsuGG

画像のアップロードの仕方が分かった。最近の家族旅行の写真を見てください。伊豆の温泉に行ったんです。http://twitpic.com/xxx

---

KenMatsuGG

これはもう1枚の画像、野球をしている息子です。息子は野球が大好きで、いつかプロの野球選手になりたいと思ってます。
http://twitpic.com/xxx

MJ_CoolChen

@KenMatsuGG いい家族写真だね。休暇の写真では、とても幸せそうに見えるよ。

KenMatsuGG

公用語を英語にするという会社の計画について、ついにうちの奥さんに話した。あんまり興味がないようだった。

KenMatsuGG
Kenji Matsumoto

MJ_CoolChen
Micheal J Chen

Sarah2000ry
Sarah Reynolds

Story 2 My Company's Official Language will be English!

@KenMatsuGG Should I start a family, or not? I'm not sure about my job situation or if I'm even ready to get married. We'll see.

> or not：もしくはそうでないか　　I'm not sure：分からない　　situation：状況
> if 〜：〜かどうか　　even 〜：〜さえ　　ready to 〜：〜する用意ができている
> get married：結婚する　　We'll see.：どうだろう、どうなるだろう

KenMatsuGG

@MJ_CoolChen You can't wait for everything to be perfect. That will never happen. Do you love your girlfriend? That's the most important.

> wait for 〜：〜を待つ　　perfect：完璧な　　happen：起こる、発生する
> girlfriend：ガールフレンド、恋人

MJ_CoolChen

@KenMatsuGG Thanks. I wish Rachel was back from her trip. It would be nice to go out to dinner, relax and forget about work for a while.

> it would be nice to 〜：〜したらいいだろう

DAY 12

KenMatsuGG

@MJ_CoolChen I decided to study with Sarah about once a week. She is very nice and can help me with English documents for work.

> once a week：週に一度　　nice：感じがいい　　document：書類

Sarah2000ry

> **MJ_CoolChen**

@KenMatsuGG 家族を作るべきかどうか？ 自分の仕事の状況がはっきりしないし、結婚する心構えができているかどうかさえ分からない。どうなるかなあ。

> **KenMatsuGG**

@MJ_CoolChen 何もかもが完ぺきになるまで待つことはできませんよ。そういうことはありません。恋人を愛しているんですか？ それが一番大切でしょう。

> **MJ_CoolChen**

@KenMatsuGG ありがとう。レイチェルが出張から戻ってくるといいんだけれど。夕食に出て、ゆっくりしてしばらく仕事のことを忘れたいな。

12日目

> **KenMatsuGG**

@MJ_CoolChen サラと週に一度くらい勉強することにしました。とても感じがよく、仕事で使う英語の書類についても助けてくれるんです。

KenMatsuGG
Kenji Matsumoto

MJ_CoolChen
Micheal J Chen

Sarah2000ry
Sarah Reynolds

Story 2　My Company's Official Language will be English!

@KenMatsuGG Don't forget to make a list of English questions that come up at work. We'll use them for next week's lesson.

> come up：生じる、発生する

MJ_CoolChen

I had a fight with my boss again today. I asked for a transfer, but he said I should stay in this department.

> have a fight：ケンカする　　ask for 〜：〜を求める、頼む　　transfer：異動
> stay：とどまる

MJ_CoolChen

I think I should either find a different position at this company or start looking for a new job. I can't stay where I am.

> either 〜 or ...：〜か…かどちらか　　position：ポジション、身分、職業
> where I am：今いる所、今ある姿

KenMatsuGG

@MJ_CoolChen It seems like you still have problems at work. That happened to me in the past when a new boss was hired.

> It seems like 〜：〜のようだ　　in the past：過去に　　hire：採用する、雇う

MJ_CoolChen

This might be a chance to work in another country. It'll be a big change, but it could be exciting.

KenMatsuGG

@MJ_CoolChen New things can be good. I'm looking forward to studying English. I feel like I'm a kid again, starting a new project.

> look forward to 〜：〜を楽しみにする　　feel like 〜：〜のような気がする
> project：計画、プロジェクト

My Company's Official Language will be English!　**Story 2**

Sarah2000ry

@KenMatsuGG 仕事で出てくる英語の疑問点リスト作るのを忘れないでね。翌週のレッスンで使うから。

MJ_CoolChen

今日、また上司とケンカした。異動を願い出たんだけれど、上司はこの部署にとどまるべきだって言うんだ。

MJ_CoolChen

この会社で別のポジションを見つけるか、新しい仕事を探すかだな。今のままではいられないよ。

KenMatsuGG

@MJ_CoolChen まだ仕事で問題があるようですね。過去に僕にもそういうことがありました、新しい上司が採用されたときに。

MJ_CoolChen

これは、ほかの国で働くためのチャンスかもしれない。大きな変化だけれど、面白いことになりそうだ。

KenMatsuGG

@MJ_CoolChen 新しいことはいいことです。僕は、英語を勉強するのを楽しみにしています。子どものころに返って、新しいことを始めるかのような気分です。

KenMatsuGG
Kenji Matsumoto

MJ_CoolChen
Micheal J Chen

Sarah2000ry
Sarah Reynolds

Story 2 My Company's Official Language will be English!

MJ_CoolChen

I can't believe this. My boss wants me to go with him to see a few customers. Why would he ask me? A test perhaps?

a few：2、3の、少数の　　customer：顧客　　perhaps：恐らく、多分

DAY 13

KenMatsuGG

I have to prepare for a client coming from the US. I'll have to explain our products in English.

client：クライアント、顧客　　US：アメリカ合衆国　　product：製品、商品

KenMatsuGG

I could ask for an interpreter, but I'll try to do it myself. Especially since English will be the company's official language.

interpreter：通訳者　　myself：自分で

MJ_CoolChen

@KenMatsuGG Rachel is back in town. I can't wait to see her. Good luck with your client meeting!

good luck with ～：～がうまくいきますように

KenMatsuGG

I might have to entertain the US visitor. I found a lot of English information on the web about Tokyo. It will be helpful for me.

entertain：もてなす、接待する　　visitor：客、訪問者　　helpful：役に立つ

MJ_CoolChen

信じられない。上司は僕に、2、3の顧客に一緒に会いに行ってほしいそうだ。なぜ僕に頼むんだろう？ もしかするとテスト？

13日目

KenMatsuGG

アメリカから来るクライアントのための準備をしなければならない。うちの製品を英語で説明しなければ。

KenMatsuGG

通訳を頼んでもいいんだけど、自分でやってみよう。特に、英語は会社の公用語になるんだから。

MJ_CoolChen

@KenMatsuGG レイチェルが街に戻っている。会うのが待ちきれないな。クライアントとの打ち合わせがうまくいくといいね！

KenMatsuGG

アメリカからのお客さんを接待しなければならないかも。東京についての英語の情報をウェブでたくさん見つけたから、それが役に立ちそうだ。

KenMatsuGG Kenji Matsumoto　　**MJ_CoolChen** Micheal J Chen　　**Sarah2000ry** Sarah Reynolds

Story 2 My Company's Official Language will be English!

KenMatsuGG

I started to make a list of English phrases that I might need to use. This is going to be hard.

hard：大変な

Sarah2000ry

@KenMatsuGG You are doing everything you can. Good luck. And don't be nervous. Just relax.

Good luck.：がんばって

MJ_CoolChen

My friend told me there have been some good real estate deals in Dubai. He says I should buy some land there.

real estate：不動産　　deal：取引　　land：土地

MJ_CoolChen

I'd consider it if I were working in the area but currently, I can't think about personal investment. I'll pass for now.

if I were working in the area：その分野で働いていたら（仮定法）
investment：投資　　pass：パスする、やめておく

DAY 14

KenMatsuGG

Practicing English expressions is tough. The expressions don't come to my mind when I try to speak.

practice：練習する　　expression：表現　　tough：大変な、難しい

KenMatsuGG

使う必要がありそうな英語表現のリストを作り始めた。これは大変そうだ。

Sarah2000ry

@KenMatsuGG あなたはできることを全部やってるわ。がんばって。緊張しないで。リラックスしてね。

MJ_CoolChen

友人が、ドバイでいい不動産取引があると言ってきた。僕に土地を買うべきだって言うんだ。

MJ_CoolChen

その分野で働いていたら考慮するだろうけど、個人投資については考えられないな。今はパスだ。

14日目

KenMatsuGG

英語の表現を練習するのは大変だな。話そうとすると、表現が浮かんでこない。

KenMatsuGG Kenji Matsumoto
MJ_CoolChen Micheal J Chen
Sarah2000ry Sarah Reynolds

Story 2 My Company's Official Language will be English!

MJ_CoolChen

@KenMatsuGG Concentrate on the ones you feel are most useful. More will come in time.

concentrate on ～：～に集中する　　in time：そのうち、時間が経つと

Sarah2000ry

@KenMatsuGG Try practicing in front of a mirror. Don't read and look at yourself when you speak. More advice at our next lesson.

mirror：鏡

KenMatsuGG

@MJ_CoolChen Thanks for the support. By the way, how are things going with your boss?

support：応援、支援

MJ_CoolChen

@KenMatsuGG Good news, the plans were cancelled. No more fights and more free time.

plan：企画、計画　　cancel：中止する、キャンセルする　　fight：ケンカ

KenMatsuGG

@MJ_CoolChen You have no reason to look for a new job anymore. I'm very happy for you.

MJ_CoolChen

@KenMatsuGG Well, I'm still looking. I checked online and am now thinking about an MBA program. That should create more opportunity.

well,：実は、あの、　　online：インターネットで
create more opportunity：機会が広がる（createは「創造する」）

MJ_CoolChen

一番役に立ちそうだと思うものに集中してみなよ。そのうちもっと出てくるようになるよ。

Sarah2000ry

鏡の前で練習してみれば。話すときに読まないで、自分のことを見るの。次のレッスンでもっとアドバイスするわ。

KenMatsuGG

@MJ_CoolChen 応援してくれてありがとう。ところで、上司とのことはどうなってるんですか？

MJ_CoolChen

@KenMatsuGG いい知らせなんだ。企画が中止になったんだ。もうケンカはなし、もっと自由時間があるんだよ。

KenMatsuGG

@MJ_CoolChen もう新しい仕事を探す理由はなくなったんだ。よかったですね。

MJ_CoolChen

@KenMatsuGG 実は、まだ探してるんだ。ネットでチェックして、MBAプログラムについて検討しているんだよ。もっとチャンスが広がるからね。

KenMatsuGG
Kenji Matsumoto

MJ_CoolChen
Micheal J Chen

Sarah2000ry
Sarah Reynolds

Story 2 My Company's Official Language will be English!

MJ_CoolChen

Rachel brought me a few things from Russia. I'm going to meet her tonight after work. We'll probably go to a local bar.

> probably：多分、恐らく　　local：地元の

DAY 15

KenMatsuGG

The US client showed up this morning to say hello. I was so nervous that I couldn't speak very fluently.

> show up：現れる、やってくる　　say hello：あいさつする　　fluently：流ちょうに

KenMatsuGG

I'll have to make more specific notes about our products. It may take a whole day to get ready.

> specific：詳細な、具体的な　　note：メモ　　a whole day：丸1日

MJ_CoolChen

Wow! MBA programs are not much cheaper online. And they take a bit longer than I thought. Worth it or not?

> not much cheaper：そんなに安くない　　longer than I thought：思ってたより長い
> (it is) worth it：それだけの価値がある

KenMatsuGG

@MJ_CoolChen Why don't you ask your co-workers? How many other people have an MBA?

> Why don't you ～?：～してはどう?　　co-worker：同僚　　how many：何人

MJ_CoolChen

レイチェルがロシアからいくつかお土産を持って来てくれた。今晩仕事の後で会うんだ。多分、地元のバーに行く。

15日目

KenMatsuGG

けさ、アメリカのクライアントがあいさつにやってきた。緊張して、あまりうまく話せなかった。

KenMatsuGG

うちの製品について、もっと詳細なメモを作らないと。準備するのに丸1日かかるぞ。

MJ_CoolChen

おっと！ MBAプログラムは、インターネットでもそんなに安くないな。それに、思ってたよりちょっと長くかかる。それだけの価値があるのかな？

KenMatsuGG

@MJ_CoolChen 同僚に聞いてみてはどう？ MBAを持っているのは、ほかに何人？

KenMatsuGG Kenji Matsumoto
MJ_CoolChen Micheal J Chen
Sarah2000ry Sarah Reynolds

Story 2 My Company's Official Language will be English!

MJ_CoolChen

@KenMatsuGG Good idea. I asked around, and a few people thought the online style is a good idea, and some said to do it overseas.

ask around：聞いて回る　　overseas：海外で

KenMatsuGG

I found a short podcast about explaining Japanese life in English, but I have no time now to listen to it.

MJ_CoolChen

I'm considering a few MBA programs here and at foreign universities. Online is still on option, longer but I could work while studying.

foreign：海外の　　university：大学　　option：選択肢

MJ_CoolChen

What will Rachel say if I do an MBA program overseas! We'll have to talk about it before I decide.

DAY 16

MJ_CoolChen

I'm going to talk about my MBA plans with Rachel at dinner tonight. I hope she understands it will be good for my career.

career：キャリア

MJ_CoolChen

@KenMatsuGG それはいい考えだ。聞いてみたら、何人かはインターネットはいい考えだって言ったけれど、海外で取った方がいいという人もいた。

KenMatsuGG

日本の生活を英語で説明するための短いポッドキャストを見つけたけれど、今は聞いている時間がないな。

MJ_CoolChen

ここでのMBAプログラムと、海外の大学でのプログラムについていくつか検討してみた。インターネットもまだ選択肢に入ってる。長めだけれど、働きながら学べるからね。

MJ_CoolChen

海外でMBAプログラムを受けるって言ったら、レイチェルは何て言うだろう！ 決める前に、話し合っておかないと。

16日目

MJ_CoolChen

今晩の夕食で、MBAのプランについて、レイチェルと話そう。僕のキャリアにとっていいことなんだって、分かってくれるといいけれど。

KenMatsuGG — Kenji Matsumoto

MJ_CoolChen — Micheal J Chen

Sarah2000ry — Sarah Reynolds

Story 2 My Company's Official Language will be English!

KenMatsuGG

I explained our products to the US clients today. It was OK, but not perfect. I think I need to be more serious about studying English.

KenMatsuGG

I've decided to take the TOEIC test again in a few months. I got a 520 score last time, but I want over 600.

last time：前回、この前　　over 〜：〜以上

MJ_CoolChen

@KenMatsuGG The TOEIC test is not so common here according to my co-workers, but that sounds like a good goal.

common：一般的な、共通の　　according to 〜：〜によると　　goal：目標

MJ_CoolChen

@KenMatsuGG I think my MBA idea will be a good goal for me, too. Targets help focus our attention and feel good when they are achieved.

target：目的、目標　　focus one's attention：意識を集中させる、焦点を絞る

KenMatsuGG

@Sarah2000ry I hope you can help me study for the TOEIC test. A score of over 600 is my new target.

Sarah2000ry

@KenMatsuGG No problem. I can bring TOEIC exercises for part of our lesson, or you can ask me about certain topics.

exercise：練習問題　　certain：特定の

KenMatsuGG
今日、アメリカのクライアントにうちの製品の説明をした。まあまあだったけど、完ぺきではない。もっとまじめに英語を勉強しないとな。

KenMatsuGG
数カ月後にまたTOEICテストを受けることにした。前回は520点だったけど、600点以上ほしいな。

MJ_CoolChen
@KenMatsuGG 僕の同僚によると、ここではTOEICテストはそんなに一般的ではないらしい。でも、いい目標だと思うよ。

MJ_CoolChen
@KenMatsuGG 僕にとっても、MBAはいい目標になると思う。目的があると集中しやすくなるし、達成したときにうれしいものね。

KenMatsuGG
@Sarah2000ry TOEICの勉強を手伝ってもらえるといいんですが。僕の新しい目標は、600点以上です。

Sarah2000ry
@KenMatsuGG 大丈夫よ。レッスンの一部としてTOEICの練習問題を持っていくわね、もしくは何か特定のテーマについて尋ねてくれてもいいわよ。

KenMatsuGG — Kenji Matsumoto
MJ_CoolChen — Micheal J Chen
Sarah2000ry — Sarah Reynolds

Story 2 My Company's Official Language will be English!

MJ_CoolChen
@KenMatsuGG Can't believe I'm really thinking about leaving the company. It really feels like home here, but the timing also feels right.

> leave the company：会社を辞める　　feel right：正しいと思える

DAY 17

KenMatsuGG
My wife wants our son to take an exam for a private junior high school. It will cost much more than public school.

> exam：試験　　private junior high school：私立中学　　public school：公立学校

MJ_CoolChen
@KenMatsuGG I went to private schools. I think the education is better. My talk with Rachel didn't go very well.

> education：教育　　go well：うまくいく

Sarah2000ry
@KenMatsuGG My friend showed me some good TOEIC materials. She's also an English teacher in Japan and gave me some good advice.

KenMatsuGG
@Sarah2000ry I am looking forward to seeing the materials. Let's make half of our next lesson for TOEIC and half for everyday conversation.

> everyday conversation：日常会話

MJ_CoolChen

@KenMatsuGG 自分が本当に会社を辞めることを考えているなんて、信じられないな。ここは自分のうちのようだけれど、タイミングがバッチリなんだ。

17日目

KenMatsuGG

うちの奥さんは、息子に私立中学を受験させたいそうだ。公立よりずっとお金がかかるな。

MJ_CoolChen

@KenMatsuGG 僕は私立の学校に行ったよ。教育の質が高いと思う。レイチェルとの話はあんまりうまくいかなかった。

Sarah2000ry

@KenMatsuGG 友達が、いいTOEICの教材を見せてくれたの。彼女も日本で英語教師をしていて、いいアドバイスをくれたわ。

KenMatsuGG

@Sarah2000ry 教材を見るのが楽しみです。次のレッスンの半分をTOEIC対策にして、残り半分を日常会話にしましょう。

KenMatsuGG Kenji Matsumoto

MJ_CoolChen Micheal J Chen

Sarah2000ry Sarah Reynolds

Story 2 My Company's Official Language will be English!

MJ_CoolChen

Rachel told me that she doesn't want me to leave Singapore. But she doesn't want to break up either.

break up：別れる

KenMatsuGG

@MJ_CoolChen You should find an MBA program in Singapore. Then you and your girlfriend can stay together. You don't want to break up, right?

KenMatsuGG

My company made an announcement about English lessons at the office. They are going to offer free lessons, once a month.

make an announcement：発表する　　free：無料の

MJ_CoolChen

My friend in Dubai suggested I buy some land there again. I think he might be connected with a real estate investor.

be connected with 〜：〜と関係がある、結び付いている　　investor：投資家

DAY 18

MJ_CoolChen

I just got some information about an American company here looking for a financial analyst. I'm going to try to get an interview.

financial analyst：金融アナリスト　　interview：面接

MJ_CoolChen

レイチェルは、僕にシンガポールを出てほしくないそうだ。でも、別れたくもない。

KenMatsuGG

@MJ_CoolChen シンガポールでMBAプログラムを見つけるべきですよ。そうしたらガールフレンドと一緒にいられるから。君も別れたくはないんでしょう？

KenMatsuGG

うちの会社が、社内で英語レッスンをやると発表した。1カ月に一度、無料レッスンを提供してくれる。

MJ_CoolChen

ドバイの友人が、また土地を買うように勧めてきた。不動産投資家とつながりがあるのかな。

18日目

MJ_CoolChen

ここで金融アナリストを探している、アメリカの会社についての情報を得た。面接を受けてみようかな。

KenMatsuGG — Kenji Matsumoto
MJ_CoolChen — Micheal J Chen
Sarah2000ry — Sarah Reynolds

Story 2 My Company's Official Language will be English!

KenMatsuGG

I decided to sign up for the TOEIC test with a group from my company. It will be good to take the test with other people.

sign up for 〜：〜に申し込む

MJ_CoolChen

The information turned out to be great. The company will be hiring in a few months, I'm excited about the possibilities.

turn out to 〜：〜だと分かる　　possibility：実現性、可能性

KenMatsuGG

The management wants all employees to get over 650 on the TOEIC, within 2 years. I don't know if I can do that.

management：経営者、管理者　　employee：従業員　　within 〜：〜以内に

Sarah2000ry

@KenMatsuGG I'm sure you can do it. Two years is a long time. Let me know what you are studying at your company classes.

let 〜 know：〜に教える、知らせる

MJ_CoolChen

@KenMatsuGG Rachel will be so happy if I get a new job here in Singapore. I'm going to see if I can also do an MBA program here.

see if 〜：〜かどうか調べる

KenMatsuGG

@MJ_CoolChen It sounds like your life is getting better. I'm happy for you. I'm just studying everyday.

KenMatsuGG

会社の団体受験で、TOEICテストに申し込むことにした。ほかの人と一緒に受験するのはいいもんだな。

MJ_CoolChen

あの情報はすごいと分かった。会社は、数カ月後に採用を行うそうだ。実現したらすごいな。

KenMatsuGG

経営者たちは、すべての従業員に、2年以内にTOEICで650点以上取ってほしいそうだ。できるかどうか分からないな。

Sarah2000ry

@KenMatsuGG あなたならきっとできるわよ。2年は長いもの。会社のクラスで何を勉強しているか教えてね。

MJ_CoolChen

@KenMatsuGG 僕がここシンガポールで新しい仕事についたら、レイチェルは大喜びだろうな。MBAプログラムもここでできるかどうか調べてみよう。

KenMatsuGG

@MJ_CoolChen 生活が順調になってきているようですね。よかった。僕は毎日ひたすら勉強しています。

KenMatsuGG — Kenji Matsumoto

MJ_CoolChen — Micheal J Chen

Sarah2000ry — Sarah Reynolds

Story 2 My Company's Official Language will be English!

MJ_CoolChen

My friend helped me arrange an interview for that new job. It's next month, but I already have a good feeling.

> arrange an interview:面接を受ける、面接を予約する
> have a good feeling:いい感触がある、いい予感がする

DAY 19

KenMatsuGG

I found out my score for the TOEIC test. I got a 580. I didn't do very well on the listening section.

> do well on 〜:〜をうまくやる、いい結果を出す

Sarah2000ry

@KenMatsuGG That's a good improvement, you should be proud. And now you know what you need to concentrate more on.

> improvement:進歩、向上   proud:自慢して、誇りに思って
> concentrate on 〜:〜に集中する

MJ_CoolChen

Big News! I'm going to start a new job. My interview yesterday went well, and I called them today to say I would take the offer.

> take an offer:申し出を受ける

MJ_CoolChen

Big News 2! I also signed up for an MBA program here and the new company will pay for it. I'll take classes at night, but it's worth it.

> sign up for 〜:〜に申し込む

MJ_CoolChen
友人が、その新しい仕事の面接を受ける手伝いをしてくれた。面接は来月だけれど、すでにいい感触がある。

19日目

KenMatsuGG
TOEICテストのスコアが来た。580点だ。リスニングがあまりよくなかったな。

Sarah2000ry
@KenMatsuGG それはすごい進歩ね、自慢していいわよ。それに今では、何にもっと集中して勉強すればいいか分かったというわけね。

MJ_CoolChen
ビッグ・ニュース！　新しい仕事を始めるんだ。昨日の面接はうまくいって、今日仕事を受けるって電話したんだよ。

MJ_CoolChen
ビッグ・ニュース２！　シンガポールでのMBAプログラムに申し込んだんだ、会社が払ってくれるんだよ！　クラスは夜になるんだけれど、その価値はあるよね。

KenMatsuGG Kenji Matsumoto

MJ_CoolChen Micheal J Chen

Sarah2000ry Sarah Reynolds

Story 2 My Company's Official Language will be English!

KenMatsuGG

@MJ_CoolChen That's great! You are going to be busy, but you will learn a lot. Good luck!

MJ_CoolChen

@KenMatsuGG Rachel is not happy that I will be busy, but she's happy I am staying in Singapore. I'm looking forward to the challenge.

KenMatsuGG

I decided to listen to an English radio program on my way to work in the morning. I'll do it everyday.

on my way to 〜：〜に行く途中で

Sarah2000ry

@KenMatsuGG Great. I think I mentioned it before, but listening to the radio is one of the best ways to improve your listening ability.

mention：言及する　　ability：能力

DAY 20

KenMatsuGG

My wife made a cake on Saturday. She said it was to celebrate my TOEIC score. I was very surprised. I didn't think she was interested.

celebrate：祝う

KenMatsuGG

@MJ_CoolChen それはすごい！ 忙しくなるね、でもたくさん学べるからね。がんばってください！

MJ_CoolChen

@KenMatsuGG レイチェルは僕が忙しくなるのがうれしくないんだけれど、シンガポールにとどまるのは喜んでいるよ。やりがいがあって楽しみだな。

KenMatsuGG

朝、通勤途中に英語のラジオ番組を聞くことにした。毎日やることにしよう。

Sarah2000ry

@KenMatsuGG よかった。前に言ったと思うけれど、ラジオを聞くのは、リスニング力を伸ばすのに最もいい方法の一つよ。

20日目

KenMatsuGG

うちの奥さんが、土曜日にケーキを焼いてくれた。僕のTOEICスコアを祝うためだって。ビックリした、興味があるとは思ってなかったから。

KenMatsuGG Kenji Matsumoto

MJ_CoolChen Micheal J Chen

Sarah2000ry Sarah Reynolds

Story 2 My Company's Official Language will be English!

MJ_CoolChen

@KenMatsuGG Congrats on the TOEIC score. I am so busy with the MBA course. I have to write a report on product development.

```
congrats on ～：～おめでとう (congratulations on ～の略)
development：開発
```

MJ_CoolChen

@KenMatsuGG Can you explain how product development works in Japan? A brief explanation is fine.

```
brief：短い、簡潔な
```

KenMatsuGG

@MJ_CoolChen I think I can give you some examples. I'm going to write it in English and email it to you.

```
give ～ example：～に例を挙げる    email：メールを送る
```

MJ_CoolChen

@KenMatsuGG Thanks. That will help me with my report. I also asked my friend from Dubai to do the same thing.

```
help ～ with …：～を…で助ける
```

KenMatsuGG

I will go to a hot spring for the spring vacation with my family. I'm really looking forward to relaxing and eating good food.

```
spring vacation：春休み
```

Sarah2000ry

@KenMatsuGG My friends and I were thinking about going to a hot spring too. Could you recommend a good place at our next lesson?

```
recommend：薦める
```

MJ_CoolChen
@KenMatsuGG　TOEICスコアおめでとう。僕はMBAのコースですごく忙しいんだ。製品開発についてのレポートを書かないといけないんだよ。

MJ_CoolChen
@KenMatsuGG　日本では製品開発はどのように行われるか教えてくれないかな？ 短い説明でいいんだ。

KenMatsuGG
@MJ_CoolChen　いくつか例を挙げられると思います。英語で書いて、メールしますね。

MJ_CoolChen
@KenMatsuGG　ありがとう。レポートを書くのに助かるよ。ドバイの友人にも、同じようにしてくれるよう頼んでいるんだ。

KenMatsuGG
春休みに家族で温泉旅行に行くんだ。ゆっくりして、おいしい物を食べるのが楽しみだな。

Sarah2000ry
@KenMatsuGG　友人と私も、温泉に行くつもりなの。次のレッスンで、いい場所を薦めてもらえないかしら？

KenMatsuGG　Kenji Matsumoto

MJ_CoolChen　Micheal J Chen

Sarah2000ry　Sarah Reynolds

Story 2　My Company's Official Language will be English!

MJ_CoolChen

I'm going to see Rachel tonight. I haven't been able to hang out with her for over three weeks. I miss her.

hang out with 〜：〜と過ごす、つきあう

DAY 21

KenMatsuGG

Studying English is tough, but it's also fun. I guess I don't have to rush. I wonder if I can get a better TOEIC score next time.

rush：急ぐ

MJ_CoolChen

@KenMatsuGG I just downloaded a new game and I'm taking a break from studying. You can't do everything at once. Pace yourself.

download：ダウンロードする　　take a break：休憩する　　at once：一度の
pace oneself：自分のペースを保つ

Sarah2000ry

@KenMatsuGG You have almost 2 years to improve. Don't worry too much. That's plenty of time and you will learn so much.

almost：ほとんど　　plenty of：たくさんの

MJ_CoolChen

My new job is good. But I might look for a new one after I finish the MBA program. I'm still considering Tokyo or Shanghai.

MJ_CoolChen
今晩レイチェルに会うんだ。3週間以上も彼女と過ごすことができなかった。彼女に会いたいな。

21日目

KenMatsuGG
英語を勉強するのは大変だけれど、楽しくもある。急がなくてもいいだろう。次のTOEICではもっといいスコアが取れるかな。

MJ_CoolChen
@KenMatsuGG 新しいゲームをダウンロードして、勉強から一休み。一度に全部はできないから。自分のペースを保たないとね。

Sarah2000ry
@KenMatsuGG 上達するのにほぼ2年あるんだから。心配しすぎないで。時間はたっぷり、たくさん学べるわよ。

MJ_CoolChen
新しい仕事は順調。でも、MBAプログラムが終わったら、新しいのを探すかもしれない。まだ東京や上海を考えているんだ。

KenMatsuGG Kenji Matsumoto **MJ_CoolChen** Micheal J Chen **Sarah2000ry** Sarah Reynolds

Story 2 My Company's Official Language will be English!

KenMatsuGG

@MJ_CoolChen It would be great to meet you if you come to Tokyo. I can show you around and introduce you to my family.

introduce 〜 to ... : 〜を…に紹介する

MJ_CoolChen

@KenMatsuGG I'd like to meet your family. I think Rachel would come with me if I got a job in Tokyo. She loves Japanese food.

Sarah2000ry

@KenMatsuGG I found an article about Japanese companies for our next lesson. Many of them are making English the official language.

KenMatsuGG

@Sarah2000ry Thanks. I'm really glad I found you on Twitter. My English is getting better thanks to you and my other friends on Twitter!

thanks to 〜 : 〜のおかげで

KenMatsuGG

@MJ_CoolChen もし君が東京に来たら、会えたらうれしいな。東京を案内して、家族にも紹介します。

MJ_CoolChen

@KenMatsuGG ぜひ君の家族に会いたいな。もし僕が東京で仕事を見つけたら、レイチェルも一緒に来ると思う。日本食が大好きだからね。

Sarah2000ry

@KenMatsuGG 次のレッスンのために、日本企業についての記事を見つけたわ。たくさんの企業が、英語を公用語にしているんですって。

KenMatsuGG

@Sarah2000ry ありがとう。あなたとツイッターで出会うことができて、本当にうれしいです。あなたとツイッター上のほかの友人たちのおかげで、僕の英語は上達しています！

KenMatsuGG — Kenji Matsumoto

MJ_CoolChen — Micheal J Chen

Sarah2000ry — Sarah Reynolds

COLUMN 2

普段の会話の中で使える表現

本書のツイートの中に、**I'm wondering 〜**や**I wonder if 〜**といった言い方がよく出てくることに気付いた方がいるかもしれません。これは「〜かなあ」という口語的な表現で、ひとりごとを言うときに使われます。このほか、普段の会話の中でよく使われる、便利な表現を紹介しましょう。

Listen! (p176)　聞いて！

What?! (p200)　何?!

What a relief! (p204)　よかった！

I guess 〜 (p48)　〜と思う

Sounds great! (p164)　よかったね！

You know what?! (p172)　知ってる?!、何だと思う?!

Why don't you 〜? (p18)　〜すれば？

What should I do (with 〜)? (p26)　（〜を）どうしよう？

Let's not 〜 (p164)　〜しないことにしよう

I'm wondering if 〜が**I'm wondering if you could give us the details.**（詳細を教えていただけないかと思うのですが）という丁寧な言い方になるように、**Sounds great!**も**It sounds great that you have been promoted.**（昇進したとは素晴らしいですね）と、応用してビジネスの場面でも使えます。基本的な表現を習得したら、状況に応じていろいろな場面でうまく使いこなせるようになるといいですね。

Story 3
We Love SuperBoys!

コンサートに行こう♪ワクワクつぶやき

Pico Toyama

Pico Toyama
@picottorin Tokyo, Japan
女性。22歳。大学を出て、ただ今フリーター生活中。日韓の混合アイドルグループSuperBoysのファン

Hi! I'm a "freeter" in Japan! I'm a big fan of "SuperBoys." It'll be great if I can find friends on Twitter. Let's talk about SuperBoys together.

ハイ! 私は日本でフリーターしてます。"スーパーボーイズ"の大ファン。ツイッターで友達が見つかったらうれしいな。一緒にスーパーボーイズの話しようよ。

SuperBoys Club
@happytimeSB Seoul
女性。韓国人。SuperBoysのファン。picottorinとライブ活動などについて情報交換をしている

A big fan of the Korean-Japanese pop group, Super-Boys. Please follow me to talk about them!

韓国・日本のポップグループ、スーパーボーイズの大ファン。彼らの話をするには、私をフォローしてね!

Jessica Wong
@rasberry08 Taipei
女性。台湾人。SuperBoysのファン仲間。日本大好き少女でもある

I'm enjoying college life here in Taiwan. I love good food, traveling, shopping, Japan, and SuperBoys.

台湾で大学生活を楽しんでいます。おいしい物と旅行、買い物、日本、スーパーボーイズが大好き。

Sona Lee
@snowwhite_ss Seoul
女性。韓国人。
SuperBoysのファン仲間

If you're interested in SuperBoys, please let me know. We can talk about them together.

スーパーボーイズに興味があるなら教えてね。一緒に話をしましょう。

Julie Watson
@juliewatson77 LA
女性。アメリカ人。
SuperBoysのファン仲間

Do you know who SuperBoys are? They are a Japanese & Korean pop group and they are quite popular in the US too! And I'm one of the biggest fans.

スーパーボーイズって知ってる? 日本人と韓国人のポップグループで、アメリカでもとっても人気があるの! 私も大ファンの一人。

Story 3 We Love SuperBoys!

DAY 1

picottorin

Hi! I'm Pico. I'm Japanese and I love SuperBoys! I want to talk about them on Twitter!

picottorin

I think I can find a lot of SuperBoys fans on Twitter because they are popular in many Asian countries!

happytimeSB

Hi, SuperBoys fans! Today they're on the Korean TV show "Music Lovers." From 8 pm on TV Seoul. Don't miss it!

be on the TV show：テレビ番組に出演する　　Korean：韓国の、韓国人の
miss：逃す

picottorin

Oh, I found the tweets on SuperBoys! There are so many. Where do I start? Who is taking about them?

tweet：つぶやき

happytimeSB

@picottorin Hello, newcomer! Let's talk about SuperBoys together. I live in Korea and I'm tweeting about their latest info!

newcomer：新人、新顔　　tweet：つぶやく、つぶやき　　latest：最新の

happytimeSB

@picottorin Who do you like the best among the SuperBoys members? My favorite is Goro from Japan!

favorite：お気に入り（の）、一番好きな

We Love SuperBoys! **Story 3**

1日目

picottorin

ハイ！ 私、ピコ。日本人で、スーパーボーイズが大好き。ツイッターで彼らのことを話したいな。

---

picottorin

ツイッターでスーパーボーイズのファンがたくさん見つかると思って。アジアのたくさんの国で人気があるからね。

---

happytimeSB

ハイ、スーパーボーイズ・ファンのみんな！ 今日彼らは韓国のテレビ番組『ミュージック・ラバーズ』に出るよ。午後8時からテレビ・ソウルで。見逃さないで！

---

picottorin

あっ、スーパーボーイズのつぶやき見つけた！ たくさんある。どこから始めよう？ 誰が話しているのかな？

---

happytimeSB

@picottorin こんにちは、新人さん！ 一緒にスーパーボーイズの話をしましょ。私は韓国に住んでいて、彼らの最新情報をつぶやいているの。

---

happytimeSB

@picottorin スーパーボーイズのメンバーの中では誰が一番好き？ 私のお気に入りは日本のゴロー！

| **picottorin** Pico Toyama | **happytimeSB** SuperBoys Club | **rasberry08** Jessica Wong | **snowwhite_ss** Sona Lee | **juliewatson77** Julie Watson |

英語Twitter多読術 **145**

Story 3 We Love SuperBoys!

picottorin

@happytimeSB Hello! I like Park Chun the best among the 3 Korean boys and I think the other 2 Japanese boys are also great.

the other：ほかの

happytimeSB

@picottorin If you add **#superboys** to your tweets, SuperBoys fans on Twitter can find you quickly. Please try it!

DAY 2

picottorin

Wow! Now I have 100 followers in just one day! So many fans on Twitter. This is amazing, thank you SuperBoys!

follower：フォロワー。ツイッター上でフォローされている相手
amazing：スゴイ、素晴らしい

picottorin

I've seen their show 5 times, and I have all of their CDs. I look for them on TV everyday. I'm following you all, send info if you have any!

I've seen：すでに見た、見たことがある (I've＝I have)
5 times：5回（「～times」で「～回」）　　follow：フォローする

happytimeSB

Hi! The SuperBoys will be on the Japanese TV show "Doki Doki Times" tonight. A famous comedian Sampeita will be too.

comedian：コメディアン、お笑い芸人

We Love SuperBoys! **Story 3**

picottorin

@happytimeSB こんにちは！ 私は三人の韓国の男の子の中で、パク・チョンが一番好き。ほかの二人の日本人の男の子もステキだと思う。

happytimeSB

@picottorin つぶやきに **#superboys** って付けると、ツイッター上のスーパーボーイズ・ファンがあなたのことをすぐに見つけられるよ。やってみて！

2日目

picottorin

わあ！ 1日で100人にフォローされてる！ ツイッターにファンがたくさんいるんだね。これってスゴイ、ありがとうスーパーボーイズ！

picottorin

コンサートは5回見たし、CDは全部持ってる。毎日テレビで探してるの。みんなフォローするから、もし何か情報があったら送ってね！

happytimeSB

ハイ！ スーパーボーイズは今晩日本のテレビ番組『ドキドキ・タイムズ』に出るよ。有名なお笑い芸人のサンペイタも一緒。

---

picottorin
Pico Toyama

happytimeSB
SuperBoys Club

rasberry08
Jessica Wong

snowwhite_ss
Sona Lee

juliewatson77
Julie Watson

picottorin

I heard SuperBoys will be in Tokyo this weekend. Please let me know if anybody knows anything about it.

weekend：週末　　let 〜 know：〜に教える、知らせる

happytimeSB

Could be. "Doki Doki Times" is a live program, so they might be staying in Tokyo over the weekend. **@picottorin** in Tokyo this weekend.

could be 〜：〜かもしれない、〜という可能性がある　　live：（テレビ番組が）生放送の
might 〜：〜かもしれない　　over the weekend：週末に

rasberry08

@picottorin Hi, I'm a fan of SuperBoys! I live in Taiwan and they're also very popular here. I have all of their CDs too!

popular：人気がある

picottorin

@rasberry08 Thanks for the reply! It's great to know that they're also popular in Taiwan. Please let me know about the fans there!

reply：返信

happytimeSB

@picottorin Did you see "Doki Doki Times" tonight? How was it? I wish I could see Japanese TV shows in Korea!

I wish I could 〜：〜できたらいいなあ

picottorin
スーパーボーイズは今週末東京にいるって聞いたけど。それについて何か知っている人がいたら教えて。

happytimeSB
かもね。『ドキドキ・タイムズ』は生番組だから、週末は東京で過ごすかも。**@picottorin** 今週末東京に

rasberry08
@picottorin ハイ！ 私、スーパーボーイズのファン！ 台湾に住んでいるんだけれど、ここでもとっても人気があるの。私もCD全部持ってる！

picottorin
@rasberry08 返信ありがとう！ 台湾でも人気があるって分かってうれしい。そっちのファンのことを教えてね！

happytimeSB
@picottorin 今晩『ドキドキ・タイムズ』見た？ どうだった？ 韓国で日本のテレビ番組が見れたらなあ！

---

picottorin — Pico Toyama
happytimeSB — SuperBoys Club
rasberry08 — Jessica Wong
snowwhite_ss — Sona Lee
juliewatson77 — Julie Watson

Story 3 We Love SuperBoys!

DAY 3

picottorin

I'm surprised. People in other countries know a lot about Japanese TV, singers and comedians! I want to learn more about Korean TV too.

---

rasberry08

@picottorin Sampeita and other comedians are sometimes on Taiwanese TV. And we know a lot of Japanese actors, such as Yamamoto Takuya!

> sometimes：ときどき　　Taiwanese：台湾の、台湾人の　　actor：俳優
> such as 〜：例えば、〜とか

happytimeSB

Oh, SuperBoys are on this month's Japanese magazine, "Oshare Girls." I hope I can buy it on the Internet.

---

happytimeSB

I have so many SuperBoys photos and I took some of them myself. Anyone interested? Check this out!
http://twitpic.com/xxx

> check 〜 out：〜を見る、調べる

happytimeSB

Actually, I've been a fan of Park Chun since he was a teenage magazine model, before he became a SuperBoys member. I can show you a pic!

> actually：実は、実際は　　teenage：十代の、ティーンの
> pic＝picture：写真、画像

We Love SuperBoys! **Story 3**

3日目

picottorin

ビックリ。よその国の人たち、日本のテレビや歌手、お笑い芸人のことをよく知ってる。私も韓国のテレビのこと、もっと知りたいな。

rasberry08

@picottorin サンペイタやほかのお笑いの人たちはときどき台湾のテレビに出てる。日本の俳優のこともたくさん知ってるよ、ヤマモト・タクヤとか！

happytimeSB

おっと、スーパーボーイズが今月、日本の雑誌『オシャレ・ガールズ』に出てる。インターネットで買えるといいんだけれど。

happytimeSB

スーパーボーイズの写真はいっぱい持ってて、自分で撮ったのもあるんだ。誰か興味ある？ これ見てね！ http://twitpic.com/xxx

happytimeSB

実は、私はパク・チョンがスーパーボーイズのメンバーになる前の、ティーン雑誌のモデルのころからのファンなの。写真見せてもいいからね！

---

picottorin Pico Toyama
happytimeSB SuperBoys Club
rasberry08 Jessica Wong
snowwhite_ss Sona Lee
juliewatson77 Julie Watson

英語Twitter多読術 **151**

Story 3 We Love SuperBoys!

picottorin

If we tweet about SuperBoys everyday from Korea, Taiwan, and Japan, they'll be more famous. What do you think?

happytimeSB

@picottorin Hey, that's a great idea! So please cover the Japanese part. I'll take care of the info from Korea.

rasberry08

@picottorin I'll tweet whenever I find info about them in Taiwan and other Taiwanese fans will do the same!

> whenever 〜：〜するときはいつでも　　do the same：同じことをする

DAY 4

picottorin

I work on weekends so I'm free during the week. I'm a part-time worker at an ice cream shop.

> during the week：平日（月〜金）の間（week endと対になる語）
> part-time worker：パート・タイマー、アルバイト

happytimeSB

I have a lot of free time now because I'm between jobs. Oh, but I have a job interview this afternoon.

> between jobs：求職中の（＝仕事の間にいる）　　job interview：仕事の面接
> this afternoon：今日の午後

We Love SuperBoys!　**Story 3**

picottorin
私たちが韓国、台湾、日本から毎日スーパーボーイズのことをつぶやいたら、彼らはもっと有名になるよね。どう思う？

happytimeSB
@picottorin　あっ、それいい考え！　じゃあ日本の情報をお願いね。私は韓国の情報を担当するから。

rasberry08
@picottorin　台湾で彼らについての情報を見つけたらいつでもつぶやくね。ほかの台湾ファンも同じようにするからね！

4日目

picottorin
私は週末働いているから、平日時間があるの。アイスクリーム・ショップでバイトしてるんだ。

happytimeSB
私は今求職中で、自由な時間がたくさんあるの。あっ、でも今日の午後は面接があるんだ。

---

picottorin Pico Toyama
happytimeSB SuperBoys Club
rasberry08 Jessica Wong
snowwhite_ss Sona Lee
juliewatson77 Julie Watson

Story 3 We Love SuperBoys!

picottorin

People like me are called "freeter" in Japan. We work at a part-time job until we find a full-time job.

full-time：フルタイムの、正規の

happytimeSB

@picottorin We have the same kind of young people here. It's really hard to find a full-time job in Korea, too.

hard：難しい

picottorin

I hope to find a good full-time job sometime, but so far I like the ice cream shop. It's "Creamy Creamy" from New York.

so far：今のところ

rasberry08

@picottorin I know "Creamy Creamy." Recently a shop opened in Taipei. They add cakes and cookies to the ice cream, don't they?

recently：最近（＝lately）　　Taipei：台北（台湾の中心都市）

happytimeSB

@picottorin I love "Creamy Creamy!" I usually go to a shop in Nandaimon and my favorite is chocolate and banana cake.

Nandaimon：南大門（韓国、ソウルにあるショッピングの名所）

happytimeSB

I'm going to record today's SuperBoys TV show because I may get home a little late tonight.

record：記録する、録画・録音する　　get home：帰宅する

picottorin
私のような人は、日本では「フリーター」って呼ばれてるの。フルタイムの仕事が見つかるまで、バイトで働くんだよ。

happytimeSB
@picottorin こっちにも同じような若者がいるよ。韓国でも、フルタイムの仕事を見つけるのは本当に難しいの。

picottorin
私もいつかいいフルタイムの仕事が見つかるといいんだけれど、今のところアイスクリーム・ショップの仕事が気に入っているの。ニューヨークから来た「クリーミー・クリーミー」だよ。

rasberry08
@picottorin 「クリーミー・クリーミー」知ってる。最近、台北にもお店ができたの。アイスクリームにケーキやクッキーを入れるんだよね。

happytimeSB
@picottorin 「クリーミー・クリーミー」大好き！ いつも南大門のお店に行くの。一番好きなのはチョコレートとバナナケーキ。

happytimeSB
今日のスーパーボーイズのテレビ番組、録画しよう。今晩は帰るのが少し遅くなるかもしれないから。

---

picottorin Pico Toyama
happytimeSB SuperBoys Club
rasberry08 Jessica Wong
snowwhite_ss Sona Lee
juliewatson77 Julie Watson

Story 3 We Love SuperBoys!

DAY 5

picottorin

I'm saving money now because I want to go to a SuperBoys concert in Korea some day. I think I need several more months to save enough.

save money：お金をためる、貯金する　　some day：いつか　　enough：十分に

picottorin

I haven't been to Korea but I'm really looking forward to visiting the country. Everyone says Seoul is a very exciting city.

look forward to 〜：〜を楽しみにする　　exciting：面白い、ワクワクするような

happytimeSB

@picottorin Please let me know when you come to Korea. I can help you buy your concert ticket.

help+（人）+（動詞の原形）：（人）が〜するのを手伝う

happytimeSB

Among Korean people, traveling to Japan is popular. Of course Tokyo is interesting but I know some people who like Hokkaido.

of course：もちろん

picottorin

@happytimeSB Hokkaido!? What is so good about Hokkaido? I didn't know that it's popular among Koreans.

happytimeSB

@picottorin We know about Hokkaido From some Korean TV dramas. It has a lot of natural beauty and good food. A nice place to travel to.

quite：とても、非常に

5日目

picottorin

いつか韓国のスーパーボーイズのコンサートに行きたいから、今お金をためてるの。十分たまるまで後何カ月かかると思う。

picottorin

また韓国に行ったことがないけど、本当に行くのを楽しみにしているの。みんな、ソウルはとっても面白い街だって言ってる。

happytimeSB

@picottorin 韓国に来るときは教えてね。コンサートのチケット買うの手伝うから。

happytimeSB

韓国人の中では、日本への旅行は人気があるんだよ。もちろん東京は面白いけれど、北海道が好きな人も知ってる。

picottorin

@happytimeSB 北海道!? 北海道の何がそんなにいいの？ 韓国人の間で人気があるなんて知らなかった。

happytimeSB

@picottorin 韓国のドラマで北海道を知ってるの。美しい自然がいっぱい、食べ物がおいしい。旅行するのにいい所だね。

---

picottorin — Pico Toyama
happytimeSB — SuperBoys Club
rasberry08 — Jessica Wong
snowwhite_ss — Sona Lee
juliewatson77 — Julie Watson

Story 3 We Love SuperBoys!

rasberry08

@picottorin I love Japan and have been to Tokyo twice. Shibuya is very interesting and I also like Ginza, it's expensive, though.

> twice：2回　　expensive：高い　　〜, though：〜だけれども

happytimeSB

The yen is very strong now, it's getting more difficult for us to visit Japan. I wish it were a little cheaper.

> the yen is strong：円が強い＝円高である　　get＋（形容詞比較級）：次第に〜する
> I wish it were 〜：〜だといいのだが　　cheap：安い

DAY 6

picottorin

It's nice to know that people around Asia are interested in Japan. What's something good about Japan? Electronic products in Akihabara?

> electronic product：電気製品

happytimeSB

I get some information through Japanese TV shows. We can watch a lot of Japanese TV dramas in Korea and we talk about them a lot.

rasberry08

For me, Japanese clothes and cosmetics are great. My friends like Japanese food such as gyudon and yakitori.

> cosmetics：化粧品

We Love SuperBoys!　**Story 3**

rasberry08

@picottorin 私、日本が大好きで、東京に２回行ったことがあるの。渋谷はとっても面白いし、銀座も好き。高いけど。

happytimeSB

今すごく円高だから、日本に行くのがどんどん難しくなっているの。もうちょっと安かったらいいんだけれど。

6日目

picottorin

アジアの人たちが、日本に興味があるって分かるのはいい感じだな。日本の何かいい物ってなんだろう？　秋葉原の電気製品とか？

happytimeSB

私は日本のテレビ番組から情報を得るんだ。韓国では日本のテレビドラマをたくさん見ることができて、それについてよく話すんだよ。

rasberry08

私にとっては、日本の洋服や化粧品がすごくいい。友達は牛丼や焼き鳥のような日本の食べ物が好きなんだって。

---

picottorin Pico Toyama
happytimeSB SuperBoys Club
rasberry08 Jessica Wong
snowwhite_ss Sona Lee
juliewatson77 Julie Watson

Story 3 We Love SuperBoys!

happytimeSB

I think Japanese cosmetics are popular because they have good quality. Rather expensive but everyone says it's worth buying.

quality：質　　rather：やや、多少　　worth -ing：～する価値がある

happytimeSB

Japanese pop singers are also quite popular here. Some of my friends study Japanese to understand the songs.

pop singer：ポップス歌手

picottorin

I only know a few Korean or Taiwanese actors. Maybe I should learn more about them, not only about the Korean members of SuperBoys!

only a few：ほんの少しだけ

picottorin

Lately, Korean singers are popular in Japan and they sing in Japanese. Their Japanese is very good, I'm so surprised!

lately：最近

snowwhite_ss

Hey, did you see the Korean magazine "K Music Fans?" There's something important about SuperBoys in it, please check it out!

We Love SuperBoys! **Story 3**

happytimeSB
日本の化粧品が人気があるのは、質がいいからだと思う。高めだけれど買う価値があるって、みんな言ってる。

happytimeSB
こっちでは、日本のポップス歌手もすごく人気があるんだ。友達の中には、歌を理解するために日本語を勉強している人もいるの。

picottorin
私は韓国や台湾の役者さんを少し知っているだけ。もっと知るようにした方がいいのかもね、スーパーボーイズの韓国人メンバーだけじゃなくて！

picottorin
最近、日本では韓国の歌手に人気があって、彼らは日本語で歌ってるの。みんな日本語上手。ビックリしちゃった！

snowwhite_ss
ねえ、韓国の雑誌『Kミュージック・ファン』見た？ スーパーボーイズについて大事なことが書いてあるの、見てみて！

---

picottorin — Pico Toyama
happytimeSB — SuperBoys Club
rasberry08 — Jessica Wong
snowwhite_ss — Sona Lee
juliewatson77 — Julie Watson

Story 3 We Love SuperBoys!

DAY 7

picottorin

I have news today! A very nice guy came to the shop and talked to me for a while. I hope he comes to my shop again!

> guy：男の人　　for a while：少しの間

rasberry08

Sounds like a chance for you. Is he good-looking? A student or a businessman? Tell me more about him if you see him again!

> sound like 〜：〜のように聞こえる、思える　　good-looking：カッコいい、ハンサム

happytimeSB

Oh, this is terrible! A Korean magazine says SuperBoys might break up. I don't believe this, I need more info.

> terrible：ひどい、とても悪い　　break up：分裂する、解散する　　believe：信じる

picottorin

That's impossible! They are going to have a big concert soon. I don't think that news is right.

> impossible：不可能、ありえない　　soon：すぐに　　right：正しい、正確な

rasberry08

I read the same story in a Taiwanese magazine. I think it's just a false rumor. If not, there would be some kind of announcement.

> false：間違っている　　rumor：うわさ　　if not：そうでなければ、
> announcement：発表

snowwhite_ss

The Korean magazine says they had an interview with their manager. So the manager wants them to break up?

We Love SuperBoys!　**Story 3**

7日目

picottorin

今日はお知らせがあるの！ とってもステキな人がお店に来て、私にちょっと話しかけていったんだ。またお店に来てくれるといいな！

rasberry08

チャンスみたいだね。カッコいい？ 学生、それともビジネスマン？ また会ったら、彼のこともっと教えてね！

happytimeSB

あーっ、これってヒドイ！ 韓国の雑誌が、スーパーボーイズが解散するかもって言ってる。信じらんない。もっと情報がいるわね。

picottorin

そんなはずないよ！ すぐに大きなコンサートがあるんだし。正確な情報だとは思えないな。

rasberry08

台湾の雑誌で同じ話を読んだよ。単なるガセネタでしょ。そうでなければ、何か発表があるはずだし。

snowwhite_ss

その韓国の雑誌は、彼らのマネジャーにインタビューしたって言ってるの。そしたらマネジャーが解散させたがってる？

| **picottorin**
Pico Toyama | **happytimeSB**
SuperBoys Club | **rasberry08**
Jessica Wong | **snowwhite_ss**
Sona Lee | **juliewatson77**
Julie Watson |

Story 3 We Love SuperBoys!

happytimeSB

Calm down everyone. It's probably nothing. Rasberry08 is right, if they were breaking up, they would let us fans know first.

> calm down：落ち着く　　probably：恐らく

happytimeSB

@picottorin Let's not worry about things that haven't happened yet. We should talk about your guy. I hope you see him again!

> let's not 〜：〜しないようにしよう　　worry：心配する　　happen：起こる
> yet：（否定文で）まだ

DAY 8

picottorin

Yes! Today he came again, while the shop manager was out. We exchanged e-mail addresses. I'm so glad, he's interested in me!

> while 〜：〜の間　　shop manager：店長　　be out：留守の、出ている
> exchange：交換する

picottorin

What should I do? Should I send him a message, or just wait until he sends me something? Someone, tell me what to do!

> or：もしくは、さもなければ　　someone：誰か

rasberry08

@picottorin Sounds great! If it was me, I would send him a message soon. He might be waiting for your message to come?

> Sounds great!：すごい！、よかった！　　wait for 〜 to ...：〜が…するのを待つ

happytimeSB
落ち着いて、みんな。多分、何でもないよ。Rasberry08の言うとおり、もし解散するんだったら、まず私たちファンに知らせるよ。

happytimeSB
@picottorin まだ起こってないことを心配するのはやめようよ。あなたのカレの話をしよう。また会えるといいね！

8日目

picottorin
やった！ 今日、店長が留守の間にまた彼が来たの。メルアド交換したんだ。よかった。彼、私に興味持ってくれてるんだ！

picottorin
どうしよう？ メールを送るべきか、彼が送ってくれるのを待つべきか？ 誰か、どうすればいいか教えて！

rasberry08
@picottorin よかったね！ 私だったら、すぐにメールを送るな。彼ってばメールが来るのを待ってるかもしれないじゃない？

---

picottorin — Pico Toyama
happytimeSB — SuperBoys Club
rasberry08 — Jessica Wong
snowwhite_ss — Sona Lee
juliewatson77 — Julie Watson

Story 3　We Love SuperBoys!

happytimeSB

@picottorin That's great! Maybe you should wait until tomorrow, and if his message doesn't come, you can just send him a short one.

> short one：短いの（＝short message。oneは代名詞）

happytimeSB

Did you see this? SuperBoys officially denied they're breaking-up in their blog. So there's nothing for us to worry about.

> officially：公式に　　deny：否定する　　blog：ブログ

picottorin

I knew it! The leader, Park Chun, always says that they really are good friends. They would not stop singing together so easily.

> I knew it!：やっぱり！分かってた！　　easily：簡単に、すぐに

rasberry08

I'm happy to hear that! I heard they're having a concert in Taipei next year. I don't want to miss it.

snowwhite_ss

I'm glad too, but I've been hearing that Goro and Park Chun are not in good moods. We should keep watching for news!

> be in (a) good mood：いい感じ、仲がいい　　keep watching：注意して見続ける

happytimeSB
@picottorin スゴイ！ 明日まで待って、カレからメールが来なかったら、短いのを送ってみれば。

happytimeSB
これ見た？スーパーボーイズがブログで公式に解散を否定しているの。だから、何も心配することはないよ。

picottorin
やっぱり！ リーダーのパク・チョンは、みんないい友達だっていつも言ってるものね。そんなに簡単に、一緒に歌うのをやめたりしないよ。

rasberry08
それはよかった！ 来年、台北でコンサートをやるって聞いたの。見逃したくないものね。

snowwhite_ss
私もうれしいけど、ゴローとパク・チョンはあんまりいい感じじゃないって聞いたよ。ニュースをチェックしてた方がいいかもね！

picottorin Pico Toyama
happytimeSB SuperBoys Club
rasberry08 Jessica Wong
snowwhite_ss Sona Lee
juliewatson77 Julie Watson

Story 3 We Love SuperBoys!

DAY 9

picottorin

Listen to this! I got a message from the guy, Kohei. He's on Twitter too, so maybe I'm going to let him know my account.

---

picottorin

Kohei already knows that I'm a big SuperBoys fan. He likes J-POP singers, such as Nishimoto Marina. That's great! I like her too.

---

rasberry08

@picottorin I know Marina! Her songs are quite popular in Taiwan and she's also a big fashion icon among Taiwanese girls.

icon：あこがれの存在、カリスマ

---

happytimeSB

Big news! SuperBoys will have a concert in the US! The first one outside Asia! Are they going to be world-famous?

outside 〜：〜の外で　　world-famous：世界的に有名な

---

picottorin

I'm so glad for them. I think their music will be loved by everyone. It will be a great success!

success：成功

---

happytimeSB

I don't have any specific information on their US concert. Is anyone in the States able to let us know what's going on?

specific：具体的な、詳しい　　the States：アメリカ（USも同様。the United States of Americaで「アメリカ合衆国」）

We Love SuperBoys! **Story 3**

9日目

picottorin

聞いて！ あの彼、コウヘイからメールがあったの。カレもツイッターやってるんだって。私のアカウント教えちゃおうかな。

---

picottorin

コウヘイはもう、私がスーパーボーイズの大ファンだって知ってるの。彼は西本マリナとか、J-POPの歌手が好きなんだって。よかった！ 私もマリナ好きだから。

---

rasberry08

@picottorin マリナ知ってる！ マリナの歌は台湾でとても人気があって、台湾の女の子の間ではファッションのカリスマでもあるんだよ。

---

happytimeSB

ビッグニュース！ スーパーボーイズがアメリカでコンサートやるんだって！ アジア以外では初めて！ 世界的に有名になっちゃうかな？

---

picottorin

よかった。彼らの音楽は、みんなに愛されるものね。きっと大成功だよ！

---

happytimeSB

アメリカ・コンサートについて、具体的な情報がないんだ。誰かアメリカに住んでいる人、どうなってるか教えてくれない？

---

picottorin
Pico Toyama

happytimeSB
SuperBoys Club

rasberry08
Jessica Wong

snowwhite_ss
Sona Lee

juliewatson77
Julie Watson

英語Twitter多読術

Story 3 We Love SuperBoys!

juliewatson77

Hi, I'm a fan of SuperBoys in LA. They're going to have a concert at LA Grand Hall in April. I will definitely go.

> LA=Los Angeles：ロサンゼルス　　definitely：絶対に、必ず

picottorin

Oh, they already have fans in America! I wish I could go to their US concert and sing their songs with everyone in English.

DAY 10

picottorin

Today I heard something depressing. The SuperBoys' latest single has not been well accepted. Is it the same in Korea and Taiwan?

> depressing：憂うつになるような、落ち込むような　　latest：最新の
> well accepted：好意的に受け入れられる、評判が良い

rasberry08

Well, I love their latest song but it's not so different from the ones before. Maybe their fans are looking for something new.

happytimeSB

I think their management is not doing a good job, and they may change their record company. That's what I heard.

> do a good job：いい仕事をする

juliewatson77

ハイ、私はロサンゼルスのスーパーボーイズ・ファン。LAグランド・ホールで4月にコンサートやるよ。ゼッタイ行く。

picottorin

わあ、もうアメリカにもファンがいるんだね！ アメリカのコンサートに行って、みんなと一緒に英語でスーパーボーイズの歌が歌えたらなあ。

10日目

picottorin

今日落ち込むことがあったの。スーパーボーイズの最新シングルが、評判よくないんだって。韓国や台湾でも同じ？

rasberry08

そうね、新しい歌は好きだけど、前のとそんなに変わらないでしょ。ファンは何か新しい物を求めているのかもね。

happytimeSB

マネジメントがよくないんだと思う。それに、レコード会社変えるかもしれないって、そう聞いたよ。

---

picottorin — Pico Toyama
happytimeSB — SuperBoys Club
rasberry08 — Jessica Wong
snowwhite_ss — Sona Lee
juliewatson77 — Julie Watson

Story 3 We Love SuperBoys!

happytimeSB

You know what?! I got a message from a big hotel chain and they said they'd like to hire me. Great! I finally got a job!

> You know what?：ねえ聞いて、知ってる？　　hire：採用する
> finally：ついに、とうとう

happytimeSB

There's one problem. If I work full-time, I won't be able to keep up with the SuperBoys latest news. That's too bad.

> keep up with ～：～ついていく　　too bad：残念な

picottorin

@happytimeSB Congratulations! Sounds like you will have an exciting job. Don't worry, we still can talk a lot about SuperBoys on Twitter!

> Congratulations!：おめでとう！

picottorin

Kohei told me to check out Marina's new music video on YouTube. It seems easy to remember. Maybe I should try it at karaoke.

> remember：覚える

rasberry08

@picottorin Did you see the Chinese signs in the video? I heard the video was shot in Taipei, but I'm not sure where it is.

> sign：看板、標識　　shot：shoot（撮影する）の過去分詞
> where it is：それがどこか

happytimeSB

何だと思う?! 大手ホテルからメールがあって、私を採用したいって。すごい! ついに仕事が見つかった!

happytimeSB

一つ問題があるんだよね。フルタイムで働くと、スーパーボーイズの最新情報についていけなくなっちゃうでしょ。残念だなあ。

picottorin

@happytimeSB おめでとう! 面白そうな仕事だね。心配しないで、ツイッターでスーパーボーイズの話がたくさんできるでしょ。

picottorin

Koheiが、マリナの新しいミュージック・ビデオをYouTubeで見てみなって。覚えやすいんだよ。カラオケで歌ってみようかな。

rasberry08

@picottorin ビデオの中国語の看板見た? このビデオ、台北で撮影されたんだって、でもどこか分からないけど。

Story 3 We Love SuperBoys!

DAY11

happytimeSB

I think I'm going to buy some new clothes for the new job. Should I get a suit? Or something more casual?

---

picottorin

@happytimeSB It's always fun to choose new clothes. You can find a lot of nice things at good prices in the Internet shops.

---

picottorin

Maybe I should think about my future more seriously. I like the ice cream shop, but I'll have to find a full-time job sooner or later.

seriously：真剣に　　sooner or later：いずれ、遅かれ早かれ

---

picottorin

Actually, I'm not sure what kind of job I really want. If I study English a little harder, maybe I'll get an interesting job?

---

rasberry08

@picottorin I think learning English helps us find a better job. I'm a student but I want to work for an international company someday.

---

picottorin

@rasberry08 One of my friends is studying through an Internet English conversation school. The teachers are in the Philippines!

English conversation school：英会話スクール　　Philippines：フィリピン

11日目

happytimeSB

新しい仕事をするのに、新しい服を買おうっと。スーツにしようかな？それとももっとカジュアルなの？

---

picottorin

@happytimeSB 新しい服を選ぶのって、いつでも楽しいよね。インターネットのショップだと、いい服がたくさんお得な値段で見つかるよ。

---

picottorin

自分の将来についてもっと真剣に考えた方がいいのかな。アイスクリーム・ショップは好きだけど、いずれフルタイムの仕事を見つけないとね。

---

picottorin

実は、本当は何の仕事をしたいか、分からないんだ。もうちょっと一生懸命英語を勉強すれば、面白い仕事ができるかな？

---

rasberry08

@picottorin 英語を勉強すると、いい仕事を見つけるのに役立つよね。私は学生だけど、いつか国際的な企業で働きたいの。

---

picottorin

@rasberry08 友達の一人は、インターネットの英会話スクールで勉強しているの。先生たちはフィリピンにいるんだよ！

| **picottorin** Pico Toyama | **happytimeSB** SuperBoys Club | **rasberry08** Jessica Wong | **snowwhite_ss** Sona Lee | **juliewatson77** Julie Watson |

Story 3　We Love SuperBoys!

happytimeSB

Hey, don't forget SuperBoys! Tonight they'll be on the Korean TV show "Extreme Music." Don't miss it!

---

picottorin

I found an interview featuring Goro in a Japanese magazine "Be Cute." He's wearing a sailor's uniform. Check it out:
http://twitpic.com/xxx

> feature 〜：〜を特集する　　sailor's uniform：水兵服

DAY 12

picottorin

Listen! Kohei asked me out to a Marina's concert. Is this a date? I'm so excited, I can hardly wait!

> Listen!：聞いて！ねえ！　　ask 〜 out to...：〜を…（デート）に誘う
> excited：すごくうれしい、ドキドキする　　hardly 〜：ほとんど〜できない

---

rasberry08

@picottorin That's fantastic! Marina's tickets are really hard to get, so I think you are extremely lucky. Enjoy your first date!

> fantastic：素晴らしい、スゴイ　　hard：難しい　　extremely：とても、極めて

---

happytimeSB

I can't believe this. The medium-sized suit didn't fit me at all. Seems like I gained some weight while I was not working.

> medium-sized：Mサイズの　　fit：（服のサイズが）合う　　not 〜 at all：全然〜でない
> Seems like 〜：〜のようだ　　gain：得る　　weight：体重

happytimeSB

ねえ、スーパーボーイズを忘れないでね！ 今晩は韓国のテレビ番組『エクストリーム・ミュージック』に出るから。見逃さないで！

picottorin

日本の雑誌『ビー・キュート』に、ゴローのインタビュー特集を見つけたよ。水兵服を着てるんだ。http://twitpic.com/xxxを見てね。

12日目

picottorin

聞いて！ コウヘイがマリナのコンサートに行こうって。これってデート？ ドキドキしちゃう。待ちきれないな！

rasberry08

@picottorin それってスゴイ！ マリナのチケットは手に入れるのが難しいんだよ、ピコは超ラッキーだね。初めてのデート、楽しんでね！

happytimeSB

これって信じらんない。Mサイズのスーツが全然入らない。仕事してない間に太ったみたい。

---

picottorin Pico Toyama
happytimeSB SuperBoys Club
rasberry08 Jessica Wong
snowwhite_ss Sona Lee
juliewatson77 Julie Watson

Story 3 We Love SuperBoys!

happytimeSB

I'm going to start jogging a little in the morning. And I should stop eating snacks between meals. Good-bye lazy life!

> jogging：ジョギング　　snack：おやつ、軽食　　meal：食事
> lazy：のんびりした、怠けた

picottorin

@happytimeSB I'm trying yoga to lose weight. It makes you relax and it's good for your health. Why don't you try it?

> lose weight：体重を減らす、やせる　　make＋(人)＋(動詞の原形)：(人)を〜させる
> relax：リラックスする、のんびりする　　Why don't you 〜？：〜してみれば？

happytimeSB

@picottorin Thanks. And I'm happy to hear that things are going well with your boyfriend. Have a great time with him!

picottorin

@happytimeSB I'm not sure if I can call him a "boyfriend" yet, but it'd be great if he is! I hope we have a wonderful time tonight!

> if 〜：〜かどうか
> if he is：もしそうだったら (if he is my boyfriendのmy boyfriendを省略)

rasberry08

@picottorin Hey, tell me how Marina's concert was. One of my friends is a big fan and she wants to know all about it.

happytimeSB

朝少しジョギングを始めようかな。食事の間にお菓子を食べるのをやめないと。のん気な生活よ、さようなら!

picottorin

@happytimeSB 私はやせるためにヨガやってるよ。リラックスするし、健康にいいんだ。happytimeSBもやってみれば?

happytimeSB

@picottorin ありがと。彼氏とうまくいっているって聞いて、私もうれしいよ。彼と一緒に楽しんできてね!

picottorin

@happytimeSB まだ彼のこと「彼氏」って呼べるかどうか分からないけど、もしそうだったらうれしいな! 今晩はステキな夜になるといいな!

rasberry08

@picottorin ねえ、マリナのコンサート、どうだったか教えてね。友達がすごいファンで、いろいろ教えてほしいって。

---

picottorin Pico Toyama
happytimeSB SuperBoys Club
rasberry08 Jessica Wong
snowwhite_ss Sona Lee
juliewatson77 Julie Watson

Story 3 We Love SuperBoys!

DAY 13

picottorin

It was perfect last night. Marina's show was great and when we said good-bye, we set up our next date. I'm so happy!

perfect：完璧な　set up：設定する

rasberry08

@picottorin I'm so glad for you. Tell me more about what he's like. Does he look like any of the SuperBoys members?

look like ～：～に似ている

picottorin

@rasberry08 He's very different than the SuperBoys, but I think he's good-looking too. He goes to college and is studying math.

different than ～：～とは違う　college：大学　math：数学

happytimeSB

@picottorin Hi, happiest girl on earth! Where are you going on your next date? Could it be a romantic place?

earth：地球

picottorin

@happytimeSB We'll meet in Shibuya, and maybe go to a movie. I'm just wondering what to wear. I don't want to be too plain, or flashy.

I'm wodering ～：～だろうかと思う　plain：地味な　flashy：派手な

13日目

picottorin

昨日の夜はカンペキだった。マリナのライブはよかったし、別れ際に、次のデートの約束したの。すっごいうれしい！

rasberry08

@picottorin それはよかったね。彼ってどんな感じか、もっと教えて。スーパーボーイズのメンバーの誰かに似てる？

picottorin

@rasberry08 スーパーボーイズとは全然違うけれど、彼もカッコいいと思うよ。大学に行ってて、数学の勉強をしているんだって。

happytimeSB

@picottorin ハイ、地上でイチバンの幸せ者！ 次のデートはどこ行くの？ ロマンチックな所だったりして？

picottorin

@happytimeSB 渋谷で会って、映画に行くかも。何を着て行こうかと思って。地味すぎず、派手すぎず、がいいんだけれど。

---

picottorin Pico Toyama
happytimeSB SuperBoys Club
rasberry08 Jessica Wong
snowwhite_ss Sona Lee
juliewatson77 Julie Watson

Story 3 We Love SuperBoys!

happytimeSB

@picottorin I heard boots are still in, so how about a short dress and long boots? I think you can find those on sale at ZARA or H&M.

```
boots：ブーツ    in：流行して    dress：ワンピース
on sale：セールになって、安くなって
```

picottorin

@happytimeSB Are those shops popular in Korea too? I've been buying clothes on the Internet lately and haven't been to those shops.

---

happytimeSB

@picottorin There are a lot of UNIQLO and MUJI shops in Korea. I like MUJI household items. They are simple but the quality is good.

```
household items：家庭用品
```

DAY 14

picottorin

Today I got my 5th message from Kohei. We're going to meet after work. I think we'll go to a barbeque restaurant.

```
after work：仕事の後で    barbeque restaurant：バーベキュー・レストラン、焼肉屋
```

picottorin

By the way, a friend of mine went to Korea last week and she bought a lot of cosmetics! In Korea, are cosmetics good and cheap?

```
by the way：ところで    last week：先週    cosmetics：化粧品
```

happytimeSB

@picottorin ブーツはまだはやってるんでしょ。短いワンピースにロングブーツは？ ZARAやH&Mのセールで見つかると思うよ。

picottorin

@happytimeSB 韓国でもそういうお店に人気があるの？ 最近インターネットで服を買っていて、そういうお店行ってなかった。

happytimeSB

@picottorin 韓国には、ユニクロや無印良品のショップがたくさんあるよ。私は、無印の家庭用品がお気に入り。シンプルだけど、質がいいよね。

14日目

picottorin

今日、コウヘイから五つ目のメッセージ。仕事の後に会うんだ。焼肉屋に行こうかな。

picottorin

ところで、友達が先週韓国に行って、化粧品をたくさん買ってきたんだ！ 韓国では、化粧品は、良くて安いの？

picottorin
Pico Toyama

happytimeSB
SuperBoys Club

rasberry08
Jessica Wong

snowwhite_ss
Sona Lee

juliewatson77
Julie Watson

happytimeSB
I heard Korean face masks are quite popular among Japanese girls. Why don't you try one, you can buy them on the Internet.

---

rasberry08
Speaking of cosmetics, I think Japanese SHISEIDO is the best in the world. Everyone around me wants SHISEIDO lately.

> speaking of 〜：〜と言えば

happytimeSB
I have SHISEIDO products! It's thought of as a luxury brand and you can be proud if you have SHISEIDO.

> product：商品、製品　It's thought of 〜：〜と考えられている
> luxury：高級な、ぜいたくな　　brand：ブランド、銘柄
> proud：自慢する、誇りに思う

---

happytimeSB
Whitening creams are really popular and a famous Korean actress is in the SHISEIDO's TV commercials for their whitening products.

> actress：女優　　TV commercial：CM

picottorin
I didn't know that SHISEIDO was so popular in other countries! So should I bring SHISEIDO cosmetics as gifts when I go to Korea or Taiwan?

> gift：贈り物、プレゼント

rasberry08
Please do! I'd love to have their sunscreen. A Japanese actress Takahashi Mayuko is on the posters for that product.

> I'd love to 〜：ぜひ〜したい　　sunscreen：日焼け止め

happytimeSB

日本の女の子の間で、韓国のフェイスマスクがすごく人気あるって聞いたよ。試してみれば、インターネットで買えるよ。

rasberry08

化粧品と言えば、日本の資生堂は世界でイチバンだと思う。最近、周りの人はみんな資生堂を欲しがってるの。

happytimeSB

資生堂の化粧品持ってる！ 高級ブランドだってされているから、資生堂持ってると自慢できるよね。

happytimeSB

ホワイトニング・クリームにすごく人気があって、韓国の有名な女優が、資生堂のホワイトニングのCMに出てる。

picottorin

資生堂がよその国でそんなに人気があるって、知らなかった！ そしたら、韓国や台湾に行くときは、資生堂の化粧品をお土産に持っていくといいの？

rasberry08

ぜひそうして！ 私、資生堂の日焼け止めが欲しいの。日本の女優の高橋マユコが、日焼け止めのポスターに出てるよね。

---

picottorin Pico Toyama

happytimeSB SuperBoys Club

rasberry08 Jessica Wong

snowwhite_ss Sona Lee

juliewatson77 Julie Watson

Story 3 We Love SuperBoys!

DAY 15

picottorin

I was really surprised to hear that plastic surgery is quite common in Korea. Doesn't it hurt? And isn't it expensive?

> plastic surgery：整形手術　　hurt：痛む

happytimeSB

It doesn't hurt at all because of the anesthesia. It might be a little expensive, so we save money for surgery.

> because of 〜：〜のために　　anesthesia：麻酔　　surgery：手術

happytimeSB

Double-edged eyelids and nose jobs are the most common. Not a big deal, just a few touches, not changing your whole face.

> double-edged eyelid：二重まぶた　　nose job：鼻の整形
> big deal：大したこと、大事なこと　　touch：手を加えること　　whole：全体の

rasberry08

At our college, everyone says we should have surgery before starting job interviews, to make us look a little better.

happytimeSB

Actually, I had a small surgery on my nose before I took this job. It cost me about 100,000 yen but I think it was worth it.

> cost：お金がかかる　　be worth it：その価値がある

picottorin

@happytimeSB Really?! Surprising! In Japan, we'd never talk about it after having that type of surgery. Wow, there's so many differences!

15日目

picottorin

韓国では整形手術がとても一般的だって聞いて、すごく驚いた。痛くないの？ それに高くない？

happytimeSB

麻酔するから全然痛くないよ。ちょっと高いかもしれないから、手術のためにお金ためるんだよ。

happytimeSB

二重まぶたと鼻の整形がイチバンよくやってるよ。大したことじゃないよ、ちょっと変えるだけで、顔がすっかり変わるわけじゃないんだから。

rasberry08

私の大学ではみんな、ちょっときれいに見せるために、就職の面接を始める前に手術を受けないとって言ってる。

happytimeSB

実は、私はこの仕事が決まる前にちょっと鼻の手術をしたんだ。10万円かかったけど、その価値はあったと思う。

picottorin

@happytimeSB 本当?! ビックリ！ 日本では、そういう手術を受けた後に、絶対そのことについて話したりしないよ。わあ、すごく違うんだね！

---

picottorin
Pico Toyama

happytimeSB
SuperBoys Club

rasberry08
Jessica Wong

snowwhite_ss
Sona Lee

juliewatson77
Julie Watson

Story 3 We Love SuperBoys!

happytimeSB

I heard there are Japanese tour groups that go to Korea to have plastic surgery. So it might be getting more common in Japan.

tour group：団体旅行のグループ

DAY 16

picottorin

Big news! SuperBoys will have a concert in Tokyo next month! How can I get a ticket? There will be such competition, chances are slim.

competition：競争　　slim：薄い

picottorin

First, I'll apply for tickets at the fan club. Only some of us can win, so I'll ask my friend to give me one if they win some.

apply for ～：～に応募する

picottorin

If I fail with the fan club, next is Internet and phone booking or convenience store booking machines. It's going to be tough.

booking：予約　　convenience store：コンビニエンス・ストア
booking machine：予約用の機械　　tough：大変な、難しい

happytimeSB

SuperBoys will be coming to Seoul after Tokyo. I'll think about how to get tickets for myself and for my friends.

happytimeSB

整形を受けるために韓国に行く、日本の団体旅行があるって聞いたよ。だから、日本でも一般的になってきているのかもね。

16日目

picottorin

ビッグ・ニュース！ スーパーボーイズが来月東京でコンサートやるって！ どうやってチケットを手に入れよう？ 競争激しくって、見込みは薄いね。

picottorin

まず、ファンクラブでチケットに応募する。当たるのは一部の人だけだから、何枚か当たったら私に１枚くれるよう友達に頼んでおく。

picottorin

もしファンクラブで手に入らなかったら次はインターネット、電話予約、コンビニのマシンでの予約。大変だぞ。

happytimeSB

スーパーボーイズは、東京の次にソウルに来るって。自分と友達のチケットをどうやって取るか、考えようっと。

---

picottorin Pico Toyama

happytimeSB SuperBoys Club

rasberry08 Jessica Wong

snowwhite_ss Sona Lee

juliewatson77 Julie Watson

happytimeSB

It's also quite difficult for us to get SuperBoys tickets. Some people buy them on Internet auctions, even though that's illegal.

auction：オークション　　even though〜：たとえ〜でも　　illegal：違法な

rasberry08

They have concerts in Tokyo and Seoul often, I wish I could go to one of those. I have to wait a year for their concert in Taipei!

often：しばしば、頻繁に

picottorin

Oh no! The fan club application limit has been met. We have to go on to the next step. OK. Who can go to a convenience store?

application：応募　　limit is met：締め切られる

rasberry08

Sounds like you have a great team. My friends say they'd go to Tokyo for a concert but it sounds impossible to get a ticket.

DAY 17

picottorin

I got one! We won tickets! I'm going to the Tokyo concert! I'm so excited. I'm going to listen to all of their CDs before the concert.

win：獲得する（過去形はwon）

happytimeSB
スーパーボーイズのチケットは、やっぱり手に入れるのが難しいんだ。インターネットのオークションで買う人もいるよ、違法だけどね。

rasberry08
スーパーボーイズは東京やソウルではよくコンサートやってるよね、私も行けたらなあ。台北のコンサートまで、1年も待たないといけないよ！

picottorin
あっ、しまった！ ファンクラブの応募、締め切られちゃった。次の手を打たないと。よし、コンビニに行けるのは誰かな？

rasberry08
いいチームがあるみたいだね。友達はコンサートのために東京に行くって言ってるけど、チケット手に入れるの無理っぽいよね。

17日目

picottorin
手に入った！ チケット取れた！ 東京コンサートに行くぞ！ ドキドキしちゃう。コンサートの前にCD全部聞いておこうっと。

---

picottorin Pico Toyama
happytimeSB SuperBoys Club
rasberry08 Jessica Wong
snowwhite_ss Sona Lee
juliewatson77 Julie Watson

Story 3 We Love SuperBoys!

rasberry08

@picottorin Wow, great! I'm so happy for you. So, are they going to sing their unreleased new song? I really want to hear it!

unreleased：未発表の、発売されていない

happytimeSB

@picottorin That's great! I heard many of the Japanese fans couldn't get tickets. They should have more shows in Tokyo.

happytimeSB

I got the information about SuperBoys' Seoul concert. But wait! I think I have to go to work on the concert day.

happytimeSB

What should I do? Can I take a day off? But I'm new in the office, I'm not sure if I can really do that.

take a day off：1日休みを取る

picottorin

@happytimeSB I think you'll be able to manage it somehow. But are you sure that you can get a ticket? How do you do that?

manage：どうにかする　　somehow：どうにかして

happytimeSB

@picottorin That's my secret! ... Just kidding, it's almost the same as what you've done. We'll try all the possible ways to get tickets.

secret：秘密　　Just kidding.：冗談です　　possible：可能な

rasberry08

@picottorin わあ、すごい！ よかったね。それで、未発売の新曲も歌うのかな？ すっごく聞きたいんだ！

happytimeSB

@picottorin それってすごい！ 日本のファンでチケット取れなかった人いっぱいいるって聞いたよ。東京公演、もっとやればいいのにね。

happytimeSB

スーパーボーイズのソウル・コンサートの情報が入ったよ！ でも、待って！ コンサートの日、私、仕事に行かないと。

happytimeSB

どうしよう？ 休めるかな？ でも、新人だから、本当にそんなことできるかどうか分かんないな。

picottorin

@happytimeSB どうにかなると思うよ、きっと。でも、チケットは本当に手に入るの？ どうやって？

happytimeSB

@picottorin それはヒミツ！ ……っていうのは冗談で、ピコがやったのとほとんど同じようなこと。チケットを手に入れるために、可能なことは全部やってみるの。

---

picottorin — Pico Toyama
happytimeSB — SuperBoys Club
rasberry08 — Jessica Wong
snowwhite_ss — Sona Lee
juliewatson77 — Julie Watson

Story 3 We Love SuperBoys!

picottorin

I'll tell Kohei about my ticket. He'll be happy for me too because he knows how hard I've been trying.

DAY 18

picottorin

I have a problem! I just realized the day of the SuperBoys' concert is Kohei's 21st birthday. I don't know what to do!

realize 〜：〜と気付く、〜と分かる

picottorin

I can't give up the concert ticket. But if I tell him that we can't spend the day together he'll be so disappointed.

give up：あきらめる　　spend：過ごす　　disappointed：がっかりする、失望する

rasberry08

@picottorin Why don't you tell him your problem honestly? He'll understand. You can celebrate his birthday on some other day.

honestly：正直に　　celebrate：祝う　　some other day：いつか別の日

happytimeSB

@picottorin In Korea, we usually have a big birthday party with a lot of friends. You don't have to spend the day by yourselves.

usually：普通、いつも　　by yourselves：あなたたち（だけ）で

picottorin

Thank you everyone! I'm going to see him tomorrow and I'll talk to him. I hope everything will be OK!

picottorin
コウヘイにチケットのこと教えてあげよ。私がどんなに努力したか知ってるから、きっと喜んでくれるよね。

18日目

picottorin
問題発生！ スーパーボーイズのコンサートの日は、コウヘイの21回目の誕生日だ。どうしよう！

picottorin
コンサートをあきらめるわけにはいかない。でも、もし誕生日を一緒に過ごすことができないって言ったら、彼、すごくガッカリするだろうなあ。

rasberry08
@picottorin 問題があるってことを、正直に話してみれば？ 分かってくれるよ。誕生日はまた別の日にお祝いできるよ。

happytimeSB
@picottorin 韓国では普通、大勢の友達と一緒に盛大なパーティーをやるんだ。自分たちだけで過ごさなくてもいいんだよ。

picottorin
みんな、ありがとう！ 明日彼に会って、話してみる。何もかもうまくいくといいんだけど！

---

picottorin Pico Toyama
happytimeSB SuperBoys Club
rasberry08 Jessica Wong
snowwhite_ss Sona Lee
juliewatson77 Julie Watson

Story 3 We Love SuperBoys!

happytimeSB

Hey, SuperBoys will be on the Japanese TV show "Music Port" to talk about their new CD and Asian concert tour. Check it out!

picottorin

I saw "Music Port"! They're going to sing their new song "Twilight Lovers" for the first time in Tokyo. What a great gift for their fans!

rasberry08

I can't go to the concert but I'm going to download their new song as soon as it's released on the Internet store.

> download：ダウンロードする　　as soon as 〜：〜したらすぐ
> release：発表する、発売する

DAY 19

picottorin

It's today! Finally, the SuperBoys concert! I'll go to the hall as soon as I finish my job at the ice cream shop. I'm getting excited!

> hall：ホール、（コンサートの）会場

rasberry08

@picottorin Great! I'm sure you'll have a great time. And what did you do about your boyfriend?

We Love SuperBoys!　**Story 3**

happytimeSB

ねえ、スーパーボーイズが日本のテレビ番組『ミュージック・ポート』に出演して、新しいCDとアジア・コンサート・ツアーについて話すんだって。見てね！

picottorin

『ミュージック・ポート』見たよ！　新曲『トワイライト・ラバーズ』を東京で最初に歌うんだって。ファンにとって素晴らしいプレゼントだね！

rasberry08

コンサートには行けないけど、新曲は、インターネットのショップで発売されたらすぐダウンロードするからね。

19日目

picottorin

今日だ！　ついに、スーパーボーイズのコンサート！　アイスクリーム・ショップの仕事が終わり次第会場に行くぞ。ドキドキしてきた！

rasberry08

@picottorin すごい！　きっと楽しめるよ。それから、彼氏のことはどうした？

---

picottorin Pico Toyama　**happytimeSB** SuperBoys Club　**rasberry08** Jessica Wong　**snowwhite_ss** Sona Lee　**juliewatson77** Julie Watson

Story 3 We Love SuperBoys!

picottorin

@rasberry08 Thanks for asking! The concert will finish at around 9, and we'll meet after that. This way, I don't have to miss anything.

> this way：この方法で、こうすると

happytimeSB

Whew! Today I'm going to start working at the hotel. Will they be nice to me? I wonder the job will be tough. I'm getting nervous!

> Whew!：ふうー、やれやれ　　nice：感じがいい　　nervous：緊張して

happpytimeSB

I heard the hotel has a lot of Japanese tourists, so maybe I should start studying Japanese.

> tourist：観光客

happytimeSB

@picottorin Hey, it'll be a great day for you! I'll be reading your tweets on my cell phone. Have a wonderful time!

> cell phone：携帯電話

picottorin

@happytimeSB Thanks, I'm on my way in to the concert. There are already so many girls around the hall. Maybe some of them are from Korea.

> on my way in to 〜：〜へ行く途中に　　already：すでに

picottorin

OK. I got a Park Chun fan, SuperBoys T-shirt, and a light stick. I'm ready for the show!

> fan：うちわ　　light stick：ペンライト
> be ready for 〜：〜の用意ができている、準備万端の

We Love SuperBoys! **Story 3**

picottorin
@rasberry08 聞いてくれてありがと！ コンサートは9時ごろ終わるから、その後会うんだ。こうすれば、何も逃さなくて済むからね。

happytimeSB
ふー！ 今日はホテルの仕事が始まる日。みんな親切かな？ 仕事は大変かな。緊張してきた！

happytimeSB
ホテルには日本の観光客がたくさんいるっていうから、日本語の勉強始めた方がいいかな。

happytimeSB
@picottorin ねえ、いい1日になりそうだね！ ピコのつぶやき、携帯で読んでるからね。楽しんできてね！

picottorin
@happytimeSB ありがと、今コンサートに行く途中。会場の周りは、もう女の子でいっぱい。韓国から来てる人もいるかもね。

picottorin
よし。パク・チョンのうちわと、スーパーボーイズのTシャツ、それからペンライトは手に入れた。ショーの準備万端だよ！

---

picottorin Pico Toyama
happytimeSB SuperBoys Club
rasberry08 Jessica Wong
snowwhite_ss Sona Lee
juliewatson77 Julie Watson

英語 Twitter 多読術

Story 3 We Love SuperBoys!

DAY 20

picottorin

What a show! I still remember everything I saw at the concert. SuperBoys are fantastic, I love them so much!

---

rasberry08

@picottorin Sounds like you had a great time. What happened with your boyfriend for his birthday celebration?

---

picottorin

@rasberry08 Please don't ask. We had a terrible fight. I haven't had a single message from him since yesterday.

fight：ケンカ　　not a single 〜：一つの〜もない

---

rasberry08

@picottorin What?! I thought you were going to celebrate his birthday after the concert. What happened? What's the reason for this?

reason：理由

---

picottorin

@rasberry08 I was late meeting him at the coffee shop. It was closed and he was waiting outside. Then all the nice bars were full.

late -ing：〜するのに遅れて　　outside：外で　　then：それから、その後
full：いっぱいの、満席の

We Love SuperBoys! **Story 3**

20日目

picottorin

すごいショー！ コンサートで見たこと、まだ全部覚えてる。スーパーボーイズはすごい。スーパーボーイズ大好き！

rasberry08

@picottorin すごく楽しかったみたいだね。彼氏との誕生日のお祝いはどうなった？

picottorin

@rasberry08 聞かないで。ひどいケンカをしたの。昨日から1通もメールが来てない。

rasberry08

@picottorin エエッ?! コンサートの後に、誕生日のお祝いをするんだと思ってた。何があったの？ 何が原因？

picottorin

@rasberry08 コーヒーショップでの待ち合わせに遅れたの。お店は閉まっちゃって、彼は外で待ってたのよ。その後、ステキなバーはみんないっぱい。

picottorin Pico Toyama	**happytimeSB** SuperBoys Club	**rasberry08** Jessica Wong	**snowwhite_ss** Sona Lee	**juliewatson77** Julie Watson

rasberry08

@picottorin Nobody was wrong then. But maybe you should say you're sorry first. He's probably waiting for your message or call.

nobody：誰も　　wrong：悪い、間違っている　　then：それなら

picottorin

@rasberry08 I can't give in so soon. If he wants to make up with me, he should call me. I just want to wait a little more.

give in：折れる、負ける　　make up with 〜：〜と仲直りする

happytimeSB

Oh no, what should I do? I made a terrible mistake at work. I erased some customer data. I don't want to go to work tomorrow!

make a mistake：間違う　　erase：消す　　customer data：顧客データ

DAY 21

picottorin

Good news! I got a message from Kohei and he apologized for what he said to me. Now I'm going to apologize too.

apologize for 〜：〜のことを謝る

We Love SuperBoys! **Story 3**

rasberry08

@picottorin それなら、誰も悪くないよ。でも、ピコがまず謝った方がいいかもね。彼はたぶん、ピコのメールか電話を待ってるよ。

picottorin

@rasberry08 そんなにすぐに謝れないよ。もし彼が私と仲直りしたいなら、電話をくれればいいんだから。もうちょっと待ってみたいの。

happytimeSB

あー、どうしよう？ 仕事で大変なミスしちゃった。顧客データを消しちゃったのよ。明日仕事に行きたくない！

21日目

picottorin

グッドニュース！ コウヘイからメールがあって、自分の言ったことを謝ったの。これから私も謝ることにする。

---

picottorin — Pico Toyama
happytimeSB — SuperBoys Club
rasberry08 — Jessica Wong
snowwhite_ss — Sona Lee
juliewatson77 — Julie Watson

Story 3 We Love SuperBoys!

picottorin

It was just a small misunderstanding. I shouldn't have been so upset. I think Kohei and I will be OK. I'm happy again!

> misunderstanding：誤解　　upset：焦る、慌てる

rasberry08

@picottorin I'm so glad for you! It would have been so sad if you two broke up because of SuperBoys. Now we can talk about SuperBoys again!

> break up：別れる

happytimeSB

What a relief! A computer engineer found the data I lost. I'm so glad, now I can concentrate on the SuperBoys concert tickets!

> What a relief!：ほっとした！安心した！　　concentrate on 〜：〜に集中する

happytimeSB

I already talked to several fans. We're going to help each other get tickets. We'll do whatever we can do to get them!

> several：何人かの　　each other：お互いに　　whatever 〜：〜することはなんでも

rasberry08

I have to wait a little longer until I can see them in Taiwan. I might buy their Tokyo or Seoul concert DVD.

picottorin

In Tokyo, the party is over and I have to get back to the real world. Should I start looking for a full-time job seriously?

> 〜 is over：〜は終わった　　real world：現実世界

We Love SuperBoys! **Story 3**

---

picottorin

ちょっとした誤解だったんだよね。そんなに焦ることなかった。コウヘイと私、大丈夫だと思う。もう立ち直ったよ！

---

rasberry08

@picottorin よかったね！ スーパーボーイズのせいで別れたなんていうことになったら、すごく悲しかったと思う。これでまたスーパーボーイズの話ができるね！

---

rasberry08

よかったー。コンピューター・エンジニアが、失くしたデータ見つけてくれた。うれしい、これでスーパーボーイズのコンサート・チケットに集中できるぞ！

---

happytimeSB

もう何人かのファンに話してみたの。お互いに協力し合ってチケットを手に入れるんだ。手に入れるためなら、できることはなんでもやるよ！

---

rasberry08

台湾でスーパーボーイズを見るまで、もうちょっと待たないと。東京かソウルのコンサートDVDでも買おうかな。

---

picottorin

東京ではもうお祭りは終わり。現実世界に戻らないと。フルタイムの仕事を本気で探し始めた方がいいかな？

| **picottorin** Pico Toyama | **happytimeSB** SuperBoys Club | **rasberry08** Jessica Wong | **snowwhite_ss** Sona Lee | **juliewatson77** Julie Watson |

Story 3 We Love SuperBoys!

picottorin

Today at the ice cream shop, I talked with some British customers in English! Thanks to Twitter, my English is improving, isn't it?

> British：イギリス人の、イギリスの　　thanks to ～：～のおかげで
> improve：上達する　　～is ..., isn't it?：～は…でしょ？

picottorin

今日、アイスクリーム・ショップで、イギリス人のお客さんと英語で話をしたの！ ツイッターのおかげで、英語が上達してたりして？

Profile

足立恵子（あだち あやこ）

東京藝術大学美術学部卒業。出版社勤務後、独立。サイクルズ・カンパニー代表。
異文化コミュニケーション・語学関連書籍と雑誌の編集者・著者・翻訳者として活躍中。
朝日カルチャーセンター 朝日JTB・交流文化塾講師、
ウェブサイト「All Aboutトラベル英会話」ガイド。
TOEIC950点、フランス語・ドイツ語・アラビア語の基礎を習得。
『英語で比べる「世界の常識」』（講談社インターナショナル）
『カラー版CD2枚付 アメリカの子どものように英語を学ぶ本』（中経出版）

ジョナサン・ナクト（Jonathan Nacht）

ニューヨーク生まれ、ニューヨーク州立大学を卒業。
コミュニティカレッジで日本語を学び、1992年に来日。
英語教師・企業内翻訳者として勤務しながら、英語学校で教材執筆・編集に携わる。
現在はフリーランスのライター・編集者・校正者、英語ウェブサイト開発コンサルタントとして活躍中。
趣味が高じ、ニューヨーク仕込みのベーグル作りのイベントなども開催している（http://ameblo.jp/jonnopan/）。
『脳を鍛える英語のクイズ—1日5分間トレーニング！』（中経出版）
『CD付 外国人がよく来るお店のやさしい英会話』（中経出版）

ツイッターアカウント：@eigo_tadoku　ハッシュタグ：#eigotadoku

140字だから楽しく読める 英語Twitter多読術
2011年4月20日　第1刷発行

著　者　　足立恵子
　　　　　ジョナサン・ナクト
発行者　　前田俊秀
発行所　　株式会社 三修社
　　　　　〒150-0001 東京都渋谷区神宮前2-2-22
　　　　　TEL：03-3405-4511
　　　　　FAX：03-3405-4522
　　　　　振替：00190-9-72758
　　　　　http://www.sanshusha.co.jp
　　　　　編集担当　安田美佳子
印刷・製本　大日本印刷株式会社

©Ayako Adachi, Jonathan Nacht 2011 Printed in Japan
ISBN978-4-384-05648-8 C2082

R〈日本複写権センター委託出版物〉
本書を無断で複写複製（コピー）することは、著作権法上の例外を除き、禁じられています。
本書をコピーされる場合は、事前に日本複写権センター（JRRC）の許諾を受けてください。
JRRC　http://www.jrrc.or.jp　　eメール：info@jrrc.or.jp　　電話：03-3401-2382

イラスト・写真：©iStockphoto.com
デザイン：深沢英次（pictex）